TIME TRAVEL

The Science and Science Fiction

Nick Redfern

ABOUT NICK REDFERN

Nick Redfern has written more than 50 books. He works full time as an author, lecturer, and journalist. He writes about a wide range of unsolved mysteries, including Bigfoot, UFOs, the Loch Ness Monster, alien encounters, and government conspiracies. His many books include *Nessie, The Real Men in Black, Shapeshifters, Monsters of the Deep, The Alien Book, The Zombie Book, The Bigfoot Book, The Monster Book, Cover-Ups & Secrets, Area 51, Secret History*, and *The Roswell UFO Conspiracy*. He writes regularly for *Mysterious Universe*. He has appeared on numerous television shows, including History Channel's *Monster Quest, Ancient Aliens*, and *UneXplained*; VH1's *Legend Hunters*; National Geographic Channel's *The Truth about UFOs* and *Paranatural*; the BBC's *Out of this World*; MSNBC's *Countdown*; and SyFy Channel's *Proof Positive*. Nick lives just a few miles from Dallas, Texas's infamous grassy knoll and can be contacted at his blog, http://nickredfernfortean.blogspot.com.

OTHER VIP BOOKS BY NICK REDFERN

ALSO FROM VISIBLE INK PRESS

"REAL NIGHTMARES" E-BOOKS BY BRAD STEIGER

PLEASE VISIT US AT VISIBLEINKPRESS.COM

Jillian Scharr is a staff writer for NBC. She, too, has carefully addressed the mind-boggling issues of time travel and wormholes and has stated: "The concept of a time machine typically conjures up images of an implausible plot device used in a few too many science-fiction storylines. But according to Albert Einstein's general theory of relativity, which explains how gravity operates in the universe, real-life time travel isn't just a vague fantasy. Traveling forward in time is an uncontroversial possibility, according to Einstein's theory. In fact, physicists have been able to send tiny particles called muons, which are similar to electrons, forward in time by manipulating the gravity around them. That's not to say the technology for sending humans 100 years into the future will be available anytime soon, though. Time travel to the past ... is even less understood. Still, astrophysicist Eric W. Davis, of the EarthTech International Institute for Advanced Studies at Austin, argues that it's possible. All you need, he says, is a wormhole, which is a theoretical passageway through space-time that is predicted by relativity."

Now let us ponder deeply on the words of Calla Cofield, whose work, according to her biography at Space.com, "has appeared in *APS News*, *Symmetry* magazine, *Scientific American*, *Nature News*, *Physics World*, and others. From 2010 to 2014 she was a producer for the *Physics Central* podcast. Previously, Calla worked at the American Museum of Natural History in New York City (hands down the best office building ever) and SLAC National Accelerator Laboratory in California." On the matter of wormholes, Cofield mentions the theoretical physicist Kip Thorne when she states in a Space.com article dated December 19, 2014:

In his 1994 publication *Black Holes and Time Warps* ... Thorne proposes a thought experiment: Say he obtains a small wormhole, which connects two points in space as if they were not separated by any distance at all. Thorne

"The wormhole theory postulates that a theoretical passage through space-time could create shortcuts for long journeys across the universe," writes Nola Taylor Redd.

takes his wormhole and puts one end in his living room, and the other aboard a spaceship parked in his front yard. Thorne's wife, Carolee, hops aboard the spaceship to prepare for a trip. The two don't have to say good-bye, though, because no matter how far away Carolee travels, they can see each other through the wormhole. They can even hold hands, as if through an open doorway.

Carolee starts up the spaceship, heads into space and travels for six hours at the speed of light. She then turns around and comes back home traveling at the same speed—a round trip of 12 hours. Thorne watches through the wormhole and sees this trip occur. He sees Carolee return from her trip, land on the front lawn, get out of the spaceship and head into the house.

But when Thorne looks out the window in his own world, his front lawn is empty. Carolee has not returned. Because she traveled at the speed of light, time slowed down for her: What was 12 hours for her was 10 years for Thorne back on Earth.

Fascinating? Indeed! We aren't over yet, though.

Bill Andrews, of *Discover* magazine, most assuredly gives us something to think about—and deeply, too. In an article dated July 30, 2019, he mentions the U.S. National Aeronautics and Space Administration (NASA) and writes: "The first problem for any explorer determined to survey a wormhole is simply finding one. While Einstein's work says they can exist, we don't currently know of any. They may actually be impossible after all, forbidden by some deeper physics that the universe obeys, but we haven't discovered. The second issue is that, despite years of research, scientists still aren't really sure how wormholes would work. Can any technology ever create and manipulate them, or are they simply a part of the universe? Do they stay open forever, or are they only traversable for a limited time? And perhaps most significantly, are they stable enough to allow for human travel? The answer to all of these: We just don't know. But that doesn't mean scientists aren't working on it. Despite the lack of actual wormholes to study, researchers can still model and test Einstein's equations. NASA's conducted legitimate wormhole research for decades, and a team described just this year how wormhole-based travel might be more feasible than previously thought."

Now I'll share with you the words of my colleague and friend at the *Mysterious Universe* website, Paul Seaburn. In 2018 Seaburn elected to immerse himself into this strange and swirling world of wormholes: "In a recent post on his *Forbes* blog, *Starts with a Bang*, theoretical astrophysicist and science writer Ethan Siegel lays out the parts and the plans for traveling backwards in time. Siegel claims this 'time machine' abides by Einstein's general theory of relativity and will not destroy the universe as we know it. Siegel proposes a sort of reverse wormhole. Instead of the conventional 'travel 40 light years out at nearly the speed of light, come back and you've aged 2 years while everyone else is 82 years older,' he proposes a wormhole with one fixed end and one that moves around at nearly the speed of light. The wormhole is created, you wait a year and then enter the end that has been in motion. When you come out at the fixed end, it's 40 years prior. That means if you entered this wormhole today, you could travel back to 1978."

> *"One amazing aspect of wormholes is that because they can behave as 'shortcuts' in space-time, they must allow for backwards time travel!"*

Finally on this particular aspect of time travel, there are the important observations of Richard F. Holman, a professor of physics at Carnegie Mellon University. He wrote about the subject in *Scientific American* on September 15, 1997: "Wormholes are solutions to the Einstein field equations for gravity that act as 'tunnels,' connecting points in space-time in such a way that the trip between the points through the wormhole could take much less time than the trip through normal space. The first wormhole-like solutions were found by studying the mathematical solution for black holes. There it was found that the solution lent itself to an extension whose geometric interpretation was that of *two* copies of the black hole geometry connected

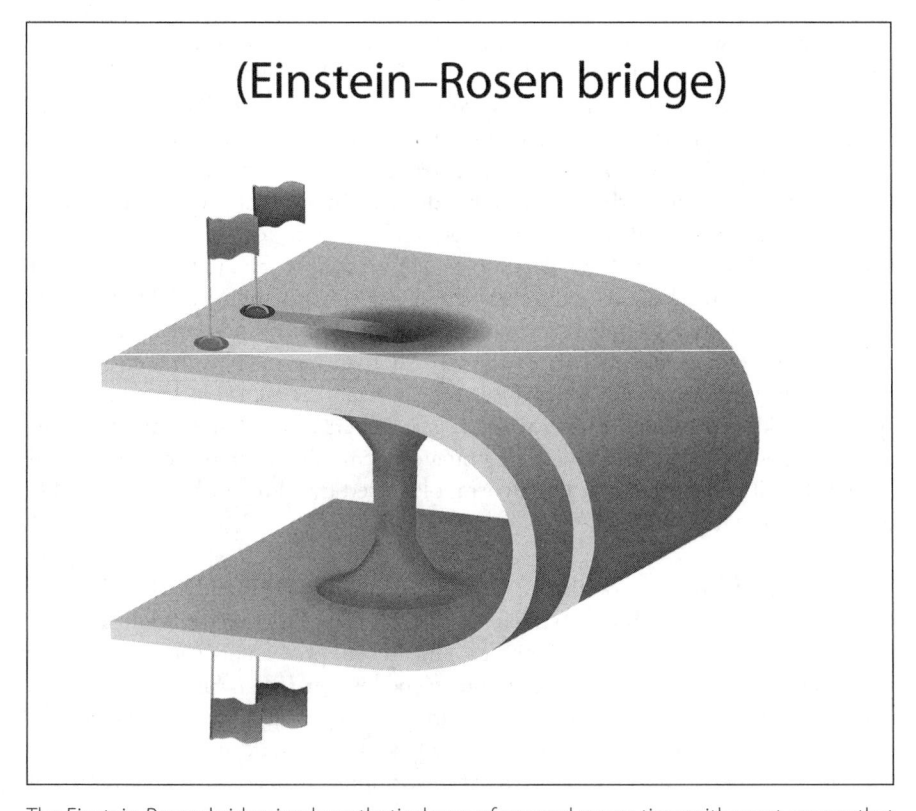

(Einstein–Rosen bridge)

The Einstein-Rosen bridge is a hypothetical area of warped space-time with great energy that can create tunnels through space-time.

by a 'throat' (known as an Einstein-Rosen bridge). The throat is a dynamical object attached to the two holes that pinches off extremely quickly into a narrow link between them. Theorists have since found other wormhole solutions; these solutions connect various types of geometry on either mouth of the wormhole. One amazing aspect of wormholes is that because they can behave as 'shortcuts' in space-time, they must allow for backwards time travel! This property goes back to the usual statement that if one could travel faster than light, that would imply that we could communicate with the past."

It's most important to note that all of those who have addressed the issue of wormholes are well-respected figures in their own, specific arenas. Moreover, they have brought to the table significant material that suggests wormholes and time travel are both a reality—something that is most important when it comes to addressing the data ahead.

And speaking of the "data ahead," how about a trip to the mysterious world of black holes?

CHAPTER 2

THE MYSTERY OF BLACK HOLES

There's no doubt that black holes are among the strangest and most mysterious phenomena in the universe. Not only that, there is an integral connection between them and the potential to travel in time. The U.S. National Aeronautics and Space Administration (NASA) explains: "A black hole is a place in space where gravity pulls so much that even light cannot get out. The gravity is so strong because matter has been squeezed into a tiny space. This can happen when a star is dying. Because no light can get out, people can't see black holes. They are invisible. Space telescopes with special tools can help find black holes. The special tools can see how stars that are very close to black holes act differently than other stars."

Gaurav Khanna, a professor of physics at the University of Massachusetts Dartmouth, provides us with some fascinating data—tying in with time travel—on black holes. At *The Conversation*, he writes in an article dated January 9, 2019: "One of the most cherished science fiction scenarios is using a black hole as a portal to another dimension or time or universe. That fantasy may be closer to reality than previously imagined. Black holes are perhaps the most mysterious objects in the universe. They are the consequence of gravity crushing a dying star without limit, leading to the formation of a true singularity—which happens when an entire star gets compressed down to a single point yielding an object with infinite density. This dense and hot singularity punches a hole in the fabric of spacetime itself, possibly opening up an opportunity for hyperspace travel. That is, a short cut through spacetime allowing for travel over cosmic scale distances in a short period."

TIME TRAVEL: THE SCIENCE AND SCIENCE FICTION

Horizons magazine has addressed this tricky, bizarre phenomenon, too. Pierre Bratschi wrote in the Swiss magazine on December 18, 2017: "Teleportation and travelling forwards through time may be possible through wormholes, the bipolar black holes that link different regions of the universe. This is the conclusion drawn from a model created by Kyriakos Papadodimas of CERN and Rik van Breukelen of the University of Geneva. Instantaneous travel and travel into the future would become possible by travelling through a wormhole, and this completely free of the time dilation predicted by Einstein's theory of relativity. It is, however, 'a purely theoretical model that would only apply to an elementary particle, such as a photon,' adds van Breukelen. With their model, the physicists have developed a new category of wormhole and describe—in theoretical terms—how information stored on a particle (e.g., using its electrical charge) could travel instantaneously to another part of space-time."

Now let's take a look at the size of black holes. As you will see, that's not an easy issue to resolve, as the following data provided by NASA makes abundantly clear: "Black holes can be big or small. Scientists think the smallest black holes are as small as just one atom. These black holes are very tiny but have the mass of a large mountain. Mass is the amount of matter, or 'stuff,' in an object. Another kind of black hole is called 'stellar.' Its mass can be up to 20 times more than the mass of the sun. There may be many, many stellar mass black holes in Earth's galaxy. Earth's galaxy is called the Milky Way. The largest black holes are called 'supermassive.' These black holes have masses that are more than 1 million suns together. Scientists have found proof that every large galaxy contains a supermassive black hole at its center. The supermassive black hole at the center of the Milky Way galaxy is called Sagittarius A. It has a mass equal to about 4 million suns and would fit inside a very large ball that could hold a few million Earths."

One of the most important questions is how, exactly, these mammoth things form in the first place. Fortunately, there are answers for us. Back to NASA: "Scientists think the smallest black holes formed when the universe began. Stellar black holes are made when the center of a very big star falls in upon itself, or collapses. When this happens, it causes a supernova. A supernova is an exploding star that blasts part of the star into space. Scientists think supermassive black holes were made at the same time as the galaxy they are in."

NASA has an important question for all of us: "If black holes are 'black,' how do scientists know they are there?"

"One of the most cherished science fiction scenarios is using a black hole as a portal to another dimension or time or universe. That fantasy may be closer to reality than previously imagined," writes physics professor Gaurav Khanna.

At its website, the organization explains: "A black hole cannot be seen because strong gravity pulls all of the light into the middle of the black hole. But scientists can see how the strong gravity affects the stars and gas around the black hole. Scientists can study stars to find out if they are flying around, or orbiting, a black hole. When a black hole and a star are close together, high-energy light is made. This kind of light cannot be seen with human eyes. Scientists use satellites and telescopes in space to see the high-energy light."

NASA adds, reassuringly: "Black holes do not go around in space eating stars, moons and planets. Earth will not fall into a black hole because no black hole is close enough to the solar system for Earth to do that. Even if a black hole the same mass as the sun were to take the place of the sun, Earth still would not fall in. The black hole would have the same gravity as the sun. Earth and the other planets would orbit the black hole as they orbit the sun now. The sun will never turn into a black

Proposing the idea of a ship circling a black hole, physicist Stephen Hawking said, "The ship and its crew would be traveling through time. Imagine they circled the black hole for five of their years. Ten years would pass elsewhere. When they got home, everyone on Earth would have aged five years more than they had."

TIME TRAVEL: THE SCIENCE AND SCIENCE FICTION

hole. The sun is not a big enough star to make a black hole. NASA is using satellites and telescopes that are traveling in space to learn more about black holes. These spacecraft help scientists answer questions about the universe."

On the issue of time travel and black holes, Elizabeth Howell at Space.com said in 2017 that it might be possible to "move a ship rapidly around a black hole, or to artificially create that condition with a huge, rotating structure. 'Around and around they'd go, experiencing just half the time of everyone far away from the black hole. The ship and its crew would be traveling through time,' physicist Stephen Hawking wrote in the *Daily Mail* in 2010. 'Imagine they circled the black hole for five of their years. Ten years would pass elsewhere. When they got home, everyone on Earth would have aged five years more than they had.' However, he added, the crew would need to travel around the speed of light for this to work. Physicist Amos Iron at the Technion-Israel Institute of Technology in Haifa, Israel pointed out another limitation if one used a machine: it might fall apart before being able to rotate that quickly."

We're getting to the point where, one day, time travel—as we understand it in movies and novels— just might become a reality. It's just a matter of ... time.

As the incredible data provided in this chapter and the previous one show, it's very clear that time travel is not just a fantasy. We're getting to the point where, one day, time travel—as we understand it in movies and novels—just might become a reality. It's just a matter of ... time.

Now let's take a look at the matter of the speed of light—something that is inextricably tied to the time travel phenomenon. The website Space.com gets right to the point: "The speed of light in a vacuum is 186,282 miles per second (299,792 kilometers per second), and in theory nothing can travel faster than light. In miles per hour, light speed is, well, a lot: about 670,616,629 mph. If you could travel at the speed of light, you could go around the Earth 7.5 times in one second.

"Early scientists, unable to perceive light's motion, thought it must travel instantaneously. Over time, however, measurements of the motion of these wave-like particles became more and more precise. Thanks to the work of Albert Einstein and others, we now understand light speed to be a theoretical limit: light speed—a constant called 'c'—is thought to be not achievable by anything with mass.... That doesn't stop sci-fi writers, and even some very serious scientists, from imagining alternative theories that would allow for some awfully fast trips around the universe."

Cathal O'Connell of *Cosmos* magazine, in an article dated April 5, 2016, provides the following: "According to Einstein's theory of special relativity, when you travel at speeds approaching the speed of light, time slows down for you relative to the outside world. This is not a just a conjecture or thought experiment—it's been measured. Using twin atomic clocks (one flown in a jet aircraft, the other stationary on Earth) physicists have shown that a flying clock ticks slower, because of its speed. In the case of the aircraft, the effect is minuscule. But if you were in a spaceship travelling at 90% of the speed of light, you'd experience time passing about 2.6 times slower than it was back on Earth. And the closer you get to the speed of light, the more extreme the time-travel."

The American Museum of Natural History features an exhibit on time machines as part of its exhibition on Albert Einstein. Related to that exhibit, the museum provides some truly mind-boggling information relative to the speed of light and time travel: "Thanks to Einstein, we know that the faster you go, the slower time passes—so a very fast spaceship is a time machine to the future. Five years on a ship traveling at 99 percent the speed of light (2.5 years out and 2.5 years back) corresponds to roughly 36 years on Earth. When the spaceship returned to Earth, the people onboard would come back 31 years in their future—but they would be only five years older than when they left. Indeed, Einstein himself could be alive today! If he could have hopped aboard a spaceship traveling at 99 percent the speed of light in 1879—the year of his birth—he would be only 17 years old upon his return to Earth today."

Now, having seen the science behind time travel and how it works, let's take a long and deep look at the phenomenon and how people from the future may already be employing it—albeit largely in stealth. Indeed, the average time traveler doesn't want to be found out—as you will soon find out.

"When you travel at speeds approaching the speed of light, time slows down for you relative to the outside world. This is not a just a conjecture or thought experiment—it's been measured," writes Cathal O'Connell in Cosmos.

CHAPTER 3

TIME TRAVEL IN THE WORLD OF ENTERTAINMENT

Theres absolutely no doubt that when it comes to how we perceive time travel, much of it is dictated by the world of entertainment: movies, television shows, and books. There's also no doubting the fact that those productions range from the intriguing to the thought-provoking, from the exciting to the disturbing, and from the unintentionally awful and laughable to the sublime. And there's no shortage of such product. Indeed, to cover just about every movie, book, and TV show on time travel would take a lifetime: the *Terminator* franchise, for example, is still rolling along nicely despite having begun in 1984. At the very least, an entire book on time travel in fiction could be written. Maybe even a second volume. I dare say a third, too. With that said, I'm going to share with you some of my favorite books, movies, and TV shows of the time-travel type.

We have to begin with the ultimate fictional classic: H. G. Wells's *The Time Machine*, first published in 1895. Certainly, novels and short stories of time travel were written before Wells took up his pen. His novella, however, remains unbeatable. It's a story of one man's journey into the future—and what a future it turns out to be. Planet Earth, in 802,701 C.E., is nothing like our world. Civilization is split into two very different groups: the Morlocks and the Eloi. The former are devolved, monstrous humans who live underground and feed on the naive and innocent Eloi, who are reliant on the Morlocks for their food and clothing. They have no understanding that they are the equivalent of cattle for the Morlocks.

TIME TRAVEL: THE SCIENCE AND SCIENCE FICTION

*That our civilization is completely gone is one of the
bleakest parts of the story. So much for the future. Such
was the success of* The Time Machine *that it was
turned into two movies.*

As the story continues, we see the final days of humanity on Earth. Our time traveler—whose name is never revealed to the reader—presses on with his treks through the millennia. That our civilization is completely gone is one of the bleakest parts of the story. Our history and society are no more. So much for the future. Such was the success of *The Time Machine* that it was turned into two movies: the first in 1960 and the second in 2002. To the credit of the producers of both movies, few changes were made to the original script.

Perhaps the world's most famous fictional time traveler is Doctor Who. The show about the time-traveling doctor first appeared on BBC television on November 23, 1963—one day after the assassination of U.S. President John F. Kennedy. Viewers quickly learned that *Doctor Who* was not, and is still not, your average time travel–themed show. For example, the doctor did not begin as a dashing, handsome, heroic type that one might expect to see in a Hollywood movie or slick television show. Quite the opposite, in fact. The eccentric doctor was played by William Hartnell, a British actor known for his work in movies and who was in his sixties when he took on the role of the now legendary time surfer.

As for the doctor's time machine, it was disguised as nothing less than a British police box. The TARDIS (short for "Time and Relative Dimension in Space"), as it is called, was no ordinary call box, however, being much larger on the inside than the outside. The white-haired old man became a star, perhaps to everyone's amazement. And the show became one of the BBC's most popular shows. Of course, Hartnell, given his age, couldn't be expected to go on playing the doctor forever. So, when he decided to quit the role in 1966, the writers came up with an ingenious idea. The doctor periodically would have to regen-

Doctor Who's time machine, the TARDIS (short for "Time and Relative Dimension in Space"), is disguised as a British police call box—but one that is much larger on the inside than on the outside.

erate, and in doing so he would take on another appearance. And another. And so on.

It should be noted that the BBC didn't have much of a budget when it originated the show. Special effects were not great, but viewers loved the quaint imagery, the plastic, rubbery monsters, and the not-so-good makeup. It was all good fun. Also loved by the fans were the archenemies of the doctor—the Daleks, the Master, and the Cybermen—and his ever-rotating companions.

Alas, all things come to an end. *Doctor Who* did so in 1989. It wasn't due to monsters or aliens, however. Ratings—rather, the lack of them—were the cause. Nevertheless, it's not easy to keep a Time Lord down for long. A movie was made about the good doctor in 1996, which was received in positive fashion. Things really changed, however, in 2005, when the doctor came back in big-time fashion. High-tech special effects caught the attention of a whole new generation, and there was barely an old man in sight. The new doctor (in all of his new regenerated forms) was cool. She still is. That's right: we now have a female Doctor Who—played by English actor Jodie Whittaker—much to the delight of the fans.

There seems to be no stopping *Doctor Who*—whether in the past, the present, or the future.

Street Eyes is a UFO-driven movie that was written and directed by Oliver Marshall. It's also a movie in which the primary character is seeking to see the future. As for the theme of the movie, you can get at least some of that from the back-cover blurb of the DVD version: "The Dead Guy roams the streets of Los Angeles possessed by an Alien. He is the key to exposing the secrets of the reptilian underworld and their plans to turn mankind into mind controlled slaves." Well, that's quite an opening salvo of words. While *Street Eyes* didn't have a huge budget and the actors are largely unknown, that doesn't take away the fact that the movie is both intriguing and thought-provoking. It's highly entertaining, too, which is always a good quality in a movie.

The primary character in the movie is Stanley. He is a UFO researcher with a fair degree of paranoia—which, as viewers soon come to see, is pretty much warranted. It's fair to say that the story encompasses numerous aspects of modern-day ufology. The movie begins in moody, atmospheric fashion out in the Nevada desert. Area 51 and a confrontation with gun-toting soldiers set the scene for what quickly follows—namely, an adventure of the ufological kind that takes some very dark and unforeseen twists.

Actress Chrissy Randall (*50 Ways to Kill Your Lover* and *9 Full Moons*) plays Stanley's girlfriend, Natalie—who we learn is pregnant. It's her pregnancy that drives much of the theme of *Street Eyes*, most of which is set in Los Angeles. The pregnancy angle is just one of many threads that weave their way through the film, keeping viewers on their toes, so to speak.

As for those threads, well, they include none other than the mysterious Men in Black. Or, rather, a Man in Black. As an interesting aside, the MIB is played by none other than UFO researcher Steve Bassett. As the story progresses, we get to see just why Natalie's baby is so important to the story. We are also introduced to one of the more controversial aspects of present-day ufology: the issue of so-called alien implants.

The cast of characters ranges from the benevolent to the downright hostile. They include "the observer" and "the dead guy." To begin with, we're not sure who are the bad guys and who are the good guys, only that Stanley and Natalie are in deep trouble. They are way over their heads, apparent pawns in an unfolding saga that threatens not just them but the freedom of the entire human race. Add to that the phenomenon of what are known as the reptilians, along with black-eyed entities, multidimensional time-traveling beings, and the "super soldiers." That's a hell of a lot of angles to cover in a movie that runs to just under one and a half hours. But Marshall and his cast and crew skillfully manage to ensure that the movie doesn't become overly complicated.

It's important, too, to note that the cast members perform well, taking on their respective characters and making them believable. The fact that much of *Street Eyes* takes place at night adds to the atmosphere and the growing threat to the primary characters, Stanley and Natalie. The movie has a good, solid ending, and—who knows?—maybe Marshall will one day treat us to a sequel. Until that happens, do check out *Street Eyes*, which has been put together by a team that was clearly enthusiastic about getting the production made and providing people with something not just to enjoy but—in terms of the overall, dark story—to ponder on, too.

The 1968 *Planet of the Apes* movie, based on Pierre Boulle's 1963 novel, brought the issue of time travel to a whole new level. It was, indeed, spectacular and a huge winner at cinemas just about everywhere. It starred Charlton Heston as astronaut George Taylor; Roddy McDowall as a chimpanzee named Cornelius; Kim Hunter as Cornelius's wife, Zira; and Maurice Evans in the role of an orangutan named Dr. Zaius.

As the movie begins, we learn that four astronauts—Taylor and his crew—are in a state of suspended animation and on a mission to a faraway star system. The team's light-speed mission from Earth to another world somehow causes distortions in the timeline, an anomaly discussed by the astronauts. Unfortunately, one of the crew, Stewart, dies on the flight. Worse still, the crew's ship plunges into a lake as it heads for the mysterious planet awaiting them. The ship quickly sinks, almost drowning the crew. Just before they are forced to flee their craft, the team realizes that around two thousand years have gone by.

The team's light-speed mission from Earth to another world somehow causes distortions in the timeline, an anomaly discussed by the astronauts.

The only option for them is to head out in search of some kind of civilization. This strange world turns out not to be too bad. In fact, the atmosphere is perfect for them. The skies are blue. It's almost as if things are too perfect. Soon the crew finds that it's not all great: On occasion the skies are filled with huge thunderclouds. Lightning illuminates the skies. The landscape is scarred and largely desert, but for the three men, it is at least survivable. All that's missing is food and water. Thankfully, they soon stumble onto a lagoon in an apparent paradise, and this is where they see, to their amazement, that there are creatures resembling humans. That is, they look just like humans, but they are primitive and lack the ability to speak. Things don't seem to be too bad at all. Matters soon change, however. It isn't long before the trio

Charlton Heston, in the 1968 film *Planet of the Apes*, plays an astronaut whose light-speed mission sends him forward not in space but in time. There he meets a human woman named Nova, played by Linda Harrison, in a society of apes.

TIME TRAVEL: THE SCIENCE AND SCIENCE FICTION

are plunged into what sounds like a deranged dream. Unfortunately, it isn't. It's a grim, horrific reality.

To their horror, the three are shocked to see this particular world is ruled by nothing less than apes. And they speak English. Humans are slaves to the apes, mere guinea pigs to Dr. Zaius, who dissects people like animals in a lab. It's not long before Taylor is the only astronaut left alive. His one comrade is a beautiful young woman named Nova. Like the other humans of this planet, she is unable to speak, but that doesn't stop the two of them from cementing a deep bond. After being experimented on, beaten, jailed, and nearly killed, Taylor knows it won't be long before his time is up. He and Nova hit the road, so to speak, without looking back. They flee to what is known as the Forbidden Zone. Dr. Zaius, tied up and unable to call for help, tells Taylor that if he continues further into that mysterious zone, he might not like what he sees. A puzzled Taylor, still with Nova, nonetheless presses on for miles and miles. The pair continue along a lengthy shoreline crashed by waves until Taylor falls to his knees. Rounding a corner on the beach, he comes across nothing less than the ruined and half-buried Statue of Liberty. To his horror, Taylor suddenly realizes that he isn't on a faraway world at all. Rather, he has been on Earth the whole time—an Earth two thousand years after his takeoff on that dicey flight. Taylor slams his fists down on the wet sand, cursing the human race for destroying its own civilization. It is one of the most memorable cinematic endings ever made.

It's intriguing to note that some of the most popular and enduring movies with a time-travel theme are not the big, adventure-driven, thriller-type productions like *12 Monkeys* and *The Philadelphia Experiment*. Rather, it's the feel-good types that never seem to go away. Moreover, they all but ensure that the whole family will be sitting together to watch. Three particular examples stand out: *It's a Wonderful Life*, *A Christmas Carol* (its alternate title being *Scrooge*), and the *Back to the Future* trilogy. Without getting too deep, I would suggest that at least part of this popularity is due to one key point: the yearning that all of us have felt at one time or another to change the past, to put things right that went wrong, to see an old friend again who passed away too early, or to spend a few hours with a long-gone, beloved pet from decades ago. Let's explore this issue a bit more.

Reviewer Robert Williams says of *Scrooge*: "This is the cream of the Christmas movie crop. The one I look forward to every year. The 1951 British version of Charles Dickens' *A Christmas Carol* will stand the test of time as the penultimate version of this tale. With flawless direction by Brian Desmond Hurst, this well-known story of a miserly Counting House owner, and the effects his mean spirit have on all those around him, come alive with the incredible acting of Alastair Sim. No one comes close to portraying the mean spiritness of Scrooge, as well as his unbounded joy upon his redemption, as well as Mr. Sim.

TIME TRAVEL: THE SCIENCE AND SCIENCE FICTION

"Noel Langley did a wonderful job of turning one of the very best Christmas books into a faithful adaptation for the screen. The 1935 British version, as well as the later American version, both lacked that indefineable something which makes any artistic endeavor worth the effort in the first place. And the movie has been done several times since, but this is the version I would choose over any other." It remains a classic to this day.

Of *It's a Wonderful Life*, renowned reviewer Roger Ebert said in 1999: "The best and worst things that ever happened to *It's a Wonderful Life* are that it fell out of copyright protection and into the shadowy no-man's-land of the public domain. Because the movie is no longer under copyright, any television station that can get its hands on a print of the movie can show it, at no cost, as often as it wants to. And that has led in the last decade to the rediscovery of Frank Capra's once-forgotten film, and its elevation into a Christmas tradition. PBS stations were the first to jump on the bandwagon in the early 1970s, using the saga of the small-town hero George Bailey as counter-programming against expensive network holiday specials. To the general amazement of TV program directors, the audience for the film grew and grew over the years, until now many families make the movie an annual ritual."

Now, let's see what has been said about the *Back to the Future* franchise. About the original 1985 film, *Wide Screenings* gives us this to muse on: "Maybe the purest theme of *Future* then would be 'having it both ways.' It doesn't take a physicist to note the absurdities such as half-faded photographs. But there is a very convincing story told here that we're dying to believe. It is a beautiful role-reversal of a child meeting his parents at the same age but able to serve as the grownup who knows better, because in this case he does."

Indeed, there is something to be said for all three of those aforementioned movies. They make us feel good. And maybe, at the end of the day, that's enough. Sometimes, cyborgs from the future, deadly pandemics, and a ruined Earth are just too much to take. Now and again, we just want something to make us feel really good.

On the matter of pandemics, there's this from *Vulture*: "When it comes to the flexibility of history, time-travel stories exist on a spectrum. On one end are stories where fate can be easily changed. *Back to the Future*, in which Marty McFly reverses all manner of poor outcomes, dwells here. Move a bit down the line and you get the *Terminator* franchise, where Armageddon is never quite stopped, but can be staved off for a bit. Farther along, you get the approach to time travel in superhero comics, where certain events—the killings of Batman's parents or Peter Parker's Uncle Ben, for example—are fixed and others can be wildly altered. Finally, move all the way to the other extreme, and you get *12 Monkeys*, where history is written in unbendable iron."

There's no doubt that in the wake of the pandemic that circled the globe in 2020, *12 Monkeys* is a movie that resonates to the worst degree possible. Don't get me wrong, it's a great movie. It's just too close to today's reality.

TIME TRAVEL: THE SCIENCE AND SCIENCE FICTION

On the matter of how the world of on-screen entertainment has embraced the controversies surrounding time travel, we encounter the world's most infamous serial killer, Jack the Ripper. Indeed, the Ripper and time surfing have gone hand in hand on several occasions—and in ingenious fashion, too. Before we get to how television productions and moviemakers have made Jack someone who has the ability to jump from century to century, it's important to understand the grisly and mysterious story that goes along with the killer himself.

In the latter part of 1888, a deadly figure roamed the shadowy, foggy back streets of Whitechapel, London, England, by night, violently slaughtering prostitutes and provoking terror throughout the entire capital. He quickly became—and remains to this day—the world's most notorious serial killer. What makes the Ripper so infamous, more than a century after his terrible crimes were committed, is that his identity remains a mystery. And everyone loves a mystery.

With that all said, let us now see how the world of Hollywood has embraced the saga of Jack the Ripper and turned the killer into nothing less than a full-blown time traveler.

Time After Time is an entertaining 1979 movie directed for Warner Brothers by Nicholas Meyer, who was also the screenwriter of the production. It stars David Warner, Mary Steenburgen, and Malcolm McDowell. It's a movie that does a good job of telling the basic story of the Ripper and his terrible killings back in London in the latter part of the nineteenth century. The story, however, has a neat twist on

Hollywood embraced the saga of Jack the Ripper and, in the movie *Time after Time*, turned the nineteenth-century serial killer into nothing less than a full-blown time traveler.

reality. None other than H. G. Wells—the brains behind the classic 1895 novel *The Time Machine*—gets entangled with the crazed murderer. We learn quickly that a friend of Wells's—John Leslie Stevenson, a surgeon—is none other than the Ripper himself. Wells, portrayed by McDowell, has no idea that his friend is the monster of London. At least, he doesn't know it at first. In an ingenious way, to avoid the police and the hangman, Jack—played by Warner—flees to twentieth-century San Francisco. Wells, too, heads off into the future, suspecting correctly that the crazed Ripper would continue his killing spree in 1979, which he does.

With the help of Steenburgen's Amy Robbins, who works at a local bank, Wells is exposed to, and shocked by, the world of the future. It's not all bad, though: the two are soon a couple. There follows a wild race to try to end Jack's killing spree. I will not spoil the ending for those who may wish to see *Time After Time* for themselves. I will say, though, that the story of a time-traveling Jack the Ripper is an entertaining and intriguing one.

Now we need to head further through time, specifically to 2017. That was when ABC made its own version of *Time After Time*, which itself was based on a 1979 novel written by Karl Alexander, who was a friend of director Nicholas Meyer. Instead of a movie of a time-traveling Jack, ABC produced a series—albeit a short one that only lasted for five of 12 filmed episodes. Writing for *IndieWire* on March 3, 2017, Liz Shannon Miller wrote: "It's weird to see such unambiguous objectification of a serial killer on screen. In fact, this might be the weirdest aspect of the time travel drama, premiering this Sunday on ABC. A reimagining, if you will, of the 1979 film written and directed by Nicholas Meyer, *Time After Time*'s first two episodes serve largely as set-up. We meet both H. G. Wells (Freddie Stroma) and his buddy Jack ([Josh] Bowman) in the late 19th century, as the two presumed friends debate what it means to live a life without fear and whether humanity will ever evolve to a better place. But the conversation gets interrupted by the arrival of the police, Jack's decision to steal the working time machine Wells was just showing off (yeah, a writer invented a working time machine, just go with it), and Wells' decision to chase after him."

Yes, the premise was very much like that of the 1979 movie. The novel, too. But it was all to no avail. Two episodes were all that ABC gave the show, and in classic Ripper fashion, the show was cut to pieces. Not even a time-traveling murderer could beat the ratings. Or, rather, the distinct lack of them.

Now it's time for me to share with you my own two all-time favorite time-travel-themed movies. One is *Millennium*, and the other is *Sound of My Voice*. Both are guaranteed to entertain and intrigue. As for the former—a 1989 film starring Kris Kristofferson, Cheryl Ladd, and Daniel Travanti—*Terrestrial Navigation* has these words for you: "Things are bad in the 30th century. The environment is severely degraded and all of humanity is barren. The people of the 30th century intend to use time travel to take people from the past who won't be missed and then send them into a far future where the Earth is (presumably) more livable." In essence, that's the theme. And it's done well, I'm pleased to say.

About the 2011 indie film *Sound of My Voice*, the website of the late Roger Ebert gives us this: "*Sound of My Voice* is a sci-fi thriller made with smoke and mirrors. No special effects, no other worlds, only the possibility of time travel, which you can't show but can only talk about. In fact, it's probably not science fiction at all, but belongs in some related category, like a story from the old *Weird Tales* magazine." If that's got you thinking, well, that's exactly what it was meant to do!

CHAPTER 4

FROM INVISIBILITY TO TIME LEAPS

When it comes to the issue of conspiracy, perhaps the biggest doozy of all is that which revolves around what is known as the Montauk Project. It's a complicated saga filled with tales of time travel (of course!), mind control, government cover-ups, secret experiments, and much more. And it's all focused around a certain facility located at Montauk Point on Long Island, New York. The story has its origins in the 1940s and involves an incredible series of classified programs run by the U.S. Navy that didn't start to surface publicly until the 1950s.

It was in 1955 that a highly controversial book on flying saucers was published. The author was Morris Ketchum Jessup, and the title of his book was *The Case for the UFO*. It was a book that, for the most part, highlighted two particular issues: first, how gravity could be harnessed and used as an energy; and second, the source of power of the mysterious flying saucers that people were seeing in the skies. It wasn't long after the book was published that Jessup was contacted by a man who wrote Jessup a number of letters that detailed something astounding. The man was one Carlos Allende, a resident of Pennsylvania.

Allende's letters were as long as they were rambling, even almost ranting, but Jessup found them oddly addictive. Allende offered what he claimed were top-secret snippets of a story that was about nothing less than invisibility. This was the type of invisibility achieved—in fictional formats, at least—in the likes of *The Invisible Man* movie of 1933, starring Claude Rains. It wasn't just invisibility that Allende had on his mind, however; it was teleportation, too, and of the kind that went drastically wrong for Jeff Goldblum's character, Seth Brundle, in 1986's *The Fly*.

Jessup read the letters with varying degrees of amazement, worry, fear, and incredulity. That's hardly surprising, given the nature of the alleged events. As Allende's tale went, at the Philadelphia Naval Yard in October 1943, the U.S. Navy reportedly managed to bring both teleportation and invisibility into the real world. According to Allende, the ship in question—the DE-173 USS *Eldridge*—vanished from Philadelphia and then very briefly appeared at Norfolk, Virginia, after which it returned to the Philadelphia Naval Yard. How did Allende know all this? He told Jessup that he was on board a ship whose crew was monitoring the experiment, the USS *Andrew Furuseth*. In a letter that detailed his own claimed sighting of the *Eldridge* vanishing from view, Allende wrote that he watched "the air all around the ship turn slightly, ever so slightly, darker than all the other air. I saw, after a few minutes, a foggy green mist arise like a cloud. I watched as thereafter the DE-173 became rapidly invisible to human eyes."

Allende's story was, to be sure, incredible. But the important question was: Was it true?

Allende told him that while the experiment worked in terms of achieving both teleportation and invisibility, it had terrible effects on the crew.

It sounded like an amazing hoax, but there was something about the story that made Jessup suspect this was not a joke at all. The more that Allende related the

At the Philadelphia Naval Yard (pictured here in 2014), the U.S. Navy reportedly managed to bring both teleportation and invisibility into the real world in October 1943.

TIME TRAVEL: THE SCIENCE AND SCIENCE FICTION

growing aspects of the tale, the more Jessup was reeled in. Allende told him that while the experiment worked in terms of achieving both teleportation and invisibility, it had terrible effects on the crew. Many of them went completely and utterly insane and lived out the rest of their lives in asylums. Some vanished from view and were never seen or heard from again. Others were fused into the deck of the ship, flesh and metal combined into one. Agonizing deaths were the only inevitabilities for these poor souls.

Jessup knew, with the potential stakes being so high, that he had to dig further into the story. Jessup was able to confirm that Allende was indeed on the *Andrew Furuseth* at the time. That was good news. Things got downright fraught for Jessup, however, when, practically out of the blue, Jessup was contacted by the U.S. Navy. They had received—anonymously—a copy of Jessup's book, *The Case for the UFO*. It was filled with scrawled messages written in pen and included numerous data on the events that had allegedly gone down in the Philadelphia Naval Yard in 1943. The Navy insisted on a meeting with Jessup. That was not good. When the meeting took place and Jessup was shown the annotated copy of his book, he was amazed to see that the annotations were the work of Carlos Allende. Jessup, worried about an official backlash, spilled the beans, revealed all he knew, then went on his way. As for the Navy, it had dozens of copies of the annotated version made. Why? No one, even to this day, is too sure. That was not quite the end of it, though. In 1959, Jessup was found dead in his car in a Florida park. For the UFO research community of the day, Jessup's death was viewed through highly suspicious eyes.

> *"The ability to put people to sleep and wake them up telepathically from a distance of a few yards to over a thousand miles became the most thoroughly tested and perfected contribution of the Soviets to international parapsychology."*

Jessup's death looked like a tragic suicide, but there are solid reasons to think that's not the answer to the mystery. It's time to share with you an extract from a Defense Intelligence Agency document from 1972 called "Controlled Offensive Behavior: USSR." This one is particularly creepy, as it focuses on the possibility of secretly using a form of extrasensory perception (ESP) to hypnotize people—including, potentially, to cause them to take their own lives. As you will see, it's a subject that elements of U.S. intelligence were worried about during the Cold War. Maybe they still are. The document tells us the following: "According to [Sheila] Ostrander and [Lynn] Schroeder, the ability to put people to sleep and wake them up telepathically from a distance of a few yards to over a thousand miles became the most thoroughly tested and perfected contribution of the Soviets to international parapsychology. It is reported that the ability to control a person's consciousness with telepathy is being further studied and tested in laboratories in Leningrad and Moscow. The work was started in the early 1920s but was not publicized until the early 1960s. The work was begun by K.O. Kotkov, a psychologist from Kharkov

University, in 1924. Kotkov could telepathically obliterate an experimental subject's consciousness from short distances or from the opposite side of town. The work was documented by [L. L.] Vasilev who conducted research of his own but could not reveal it under Stalin's regime. The reality of telepathic sleep-wake, backed by columns of data, might be the most astonishing part of Vasilev's experiments in mental suggestion."

Now let's see what else the United States learned: "Parapsychologists in Leningrad and Moscow are involved in the telepathic manipulation of consciousness, now recording successes with the EEG. Doctor V. Raikov is involved in this EEG research as well as E. Naumov. Naumov reports that mental telepathy woke up a hypnotized subject (by telepathy) six of eight times. Naumov remarked that as soon as the telepathic 'wake up' is sent, trance becomes less and less deep, full consciousness returning in twenty to thirty seconds. In the Leningrad laboratory of Doctor Paul Gulyaiev (Bekhterev Brain Institute), friends of subjects have been trained to put them to sleep telepathically."

The following question was asked within the Pentagon: "Why are the Soviets again hard at work on the telepathic control of consciousness?" The Defense Intelligence Agency (DIA) responded: "Doctor I. Kogan, like Vasilev, is probably doing it for theoretical reasons; still trying mathematically to prove that an electromagnetic carrier of telepathy is possible. Why other scientists may be delving into control of consciousness by ESP is another question. During telepathic sleep is an individual simply dreaming his own private dreams or does someone else hold sway? The cur-

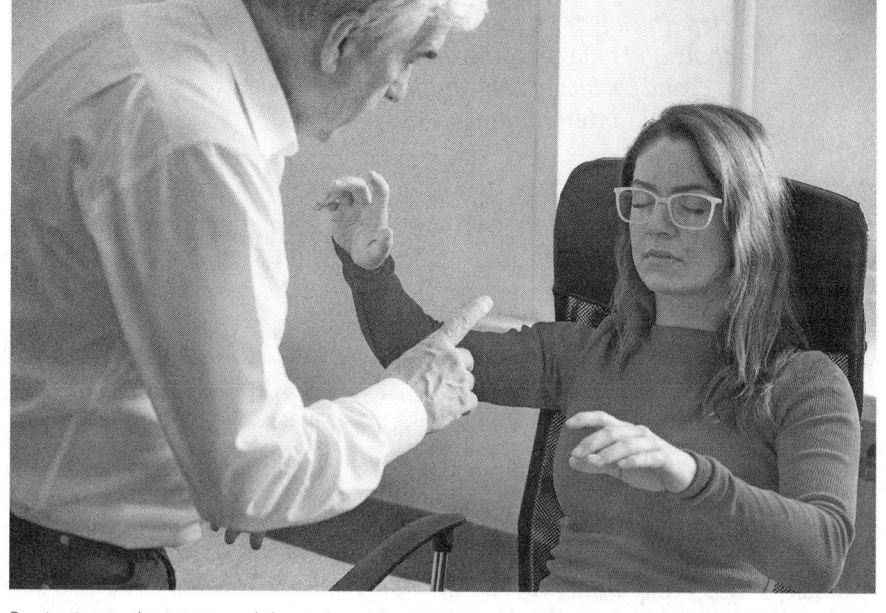

Beginning in the 1920s and throughout the Cold War, the Soviets reportedly conducted experiments with hypnotism, telepathy, and other forms of extrasensory perception.

rent Soviets have not divulged the psychological details about their telepathic manipulation of consciousness. Vasilev describes some revelations in his book but little else has been reported. Doctor Stefan Manczarski of Poland predicts that this new field of telepathy will open up new avenues for spreading propaganda. He feels that the electromagnetic theory is valid and believes, therefore that telepathy can be amplified like radio waves. Telepathy would then become a subtle new modus for the 'influences' of the world. Doctor Manczarski's wave ideas are still very debatable but what about telepathy someday becoming a tool for influencing people?"

The DIA added: "Hypnotizing someone telepathically probably comes over as a more eerie, mystifying, almost diabolical act in the US than it does in the Soviet Union. The US is really just becoming adjusted to some of the aspects of hypnotism. Since the turn of the century, the Soviets have been exploring and perfecting the various advantages that hypnotism provides. In the Soviet Union hypnotism is a common tool like X-rays, used in medicine, psychotherapy physiology, psychology, and experimental pedagogy. The Soviets have been reportedly working on the effects of drugs used in combination with psychic tests. Vasilev used mescarine in the early days and more recently M.S. Smirnov, of the Laboratory of Vision, Institute of Problems of Information Transmission of the USSR Academy of Science, has been obtaining psychic success with psilocybin. The tests that Vasilev had perfected may have a more interesting future in them than the developer had imagined. Manipulating someone else's consciousness with telepathy, guiding him in trance colorful uses are too easy to conjure. The ability to focus a mental whammy on an enemy through hypnotic telepathy has surely occurred to the Soviets."

And, finally, we have this: "Visiting Soviet psi labs in 1967, Doctor Ryzl says he was told by a Soviet, 'When suitable means of propaganda are cleverly used, it is possible to mold any man's conscience so that in the end he may misuse his abilities while remaining convinced that he is serving an honest purpose.' Ryzl continues, 'The USSR has the means to keep the results of such research secret from the rest of the world and, as practical applications of these results become possible, there is no doubt that the Soviet Union will do so.' What will ESP be used for? 'To make money, *and as a weapon* [italics mine],' Ryzl states flatly."

If such possibilities existed in 1972, that begs the question: Is ESP still being used to hypnotize people unknowingly? It's a disturbing question. So, too, could be the answer.

In the late 1970s, the story of the incident at Philadelphia was picked up again, this time by researchers Bill Moore and the late Charles Berlitz. The result: their 1979 book *The Philadelphia Experiment*. One of the more interesting things that the pair uncovered was a newspaper clipping headlined "Strange Circumstances Surround Tavern Brawl." It reads as follows: "Several city police officers responding to a call to aid members of the Navy Shore Patrol in breaking up a tavern brawl near the U.S. Navy docks here last night got something of a surprise when they arrived on the scene to find the place empty of customers. According to a pair of very nervous waitresses, the Shore Patrol had arrived first and cleared the place out—but not before two of the sailors involved allegedly did a disappearing act. 'They just sort of vanished into thin air … right there,' reported one of the frightened hostesses, 'and I ain't been drinking either!' At that point, according to her account, the Shore Patrol proceeded to hustle everybody out of the place in short order."

The clipping continues: "A subsequent chat with the local police precinct left no doubts as to the fact that some sort of general brawl had indeed occurred in the vicinity of the dockyards at about eleven o'clock last night, but neither confirmation nor denial of the stranger aspects of the story could be immediately obtained. One reported witness succinctly summed up the affair by dismissing it as nothing more than 'a lot of hooey from them daffy dames down there,' who, he went on to say, were probably just looking for some free publicity. Damage to the tavern was estimated to be in the vicinity of six hundred dollars.'"

While the story is certainly a controversial one, in the 1990s it was given a degree of support thanks to a man named George Mayerchak. For a period of time in 1949, Mayerchak—a sailor—was a patient at the Philadelphia Navy Hospital, getting over a bad case of pneumonia. It was while Mayerchak was in the hospital that he heard very weird tales of the top-secret experiment that had occurred six years earlier. Tales of the vanishing sailors and the invisible ship abounded, as did the story of the barroom brawl and the men who disappeared into states of nothingness. Mayerchak said, though, that rather than completely vanishing, they "flickered" on and off, like a light bulb—which surely would have been a bizarre thing to see.

Further amazing testimony came from Harry Euton. He confided in Bill Moore that with his "top secret" clearance during World War II, Euton was directly involved in the highly classified experiment. Reportedly, it was an experiment designed to shield U.S. ships from detection by Nazi radar systems. Something went wrong, though, explained Euton, who said that the ship became invisible. As he looked down and couldn't see any sign of the ship, Euton instantly felt nauseous and reached out for a nearby cable that he knew was there and could feel but couldn't see. Euton too confirmed that several of the men vanished, never to be seen again, and that the surviving crew didn't look as they did normally—curious words that Euton preferred not to expand upon. All of which brings us to Montauk.

As the story goes, because World War II was at its height when the Philadelphia Experiment occurred, and because no one fully understood how terribly wrong the experiment had gone, a decision was made to put the whole thing on hold until after hostilities with the Nazis were over and normality had returned to the world. It was, Montauk researchers say, in 1952 that tentative steps were taken to resurrect the Philadelphia Experiment for a whole new team of scientists. Supposedly, though, the U.S. Congress, fearful of opening what may have been a definitive Pandora's box, got cold feet and axed the program. That didn't end matters, however. The U.S. military was determined to push on and funded the classified program in a very alternative way: by using a massive stash of gold that had been secured from the Germans when the war came to its end in 1945.

The money was now available. The scientific team was eager and ready to go. The beginnings of the Montauk were about to launch. Reportedly, things began at the Brookhaven National Laboratory, situated on Long Island and under the control of the Atomic Energy Commission; it was later taken over by a powerful and shadowy elite that worked out of the Montauk Air Force Station, also on Long Island. Montauk investigators allege that today, the research into time travel, invisibility, mind control, and much more is still going on—not exactly at the old base but a hundred feet below it, in fortified bunkers. How much of this can be confirmed? Can any of it be confirmed? These are important and crucial questions.

There is no doubt that such a military facility did exist on Long Island. Even as long ago as the latter part of the eighteenth century, the area was noted for its ability to provide the military with the perfect lookout spot for potential enemy navies attempting to invade via the waters of the Atlantic Ocean, such as the Brits at the height

A resurrection of the infamous Philadelphia Experiment reportedly was launched at Brookhaven National Laboratory (pictured here), on Long Island, and then was taken over by researchers at Montauk Air Force Station, also on Long Island.

TIME TRAVEL: THE SCIENCE AND SCIENCE FICTION

of the War of Independence. In World War I, the military was using the area to keep watch for any and all potential German troops who might try to launch an assault. It was in 1942, however, that things really began to take off. It was a direct result of the terrible attack on Pearl Harbor, Hawaii, by the Japanese in December 1941 that plans were initiated to create what was initially termed Camp Hero. It was ingeniously camouflaged as a pleasant little fishing port. In reality, it was one of the most strategically positioned military facilities in the entire country. When the war was over, the base became largely inoperative—until, that is, it became clear that the Soviets were going to be the next big threat. Camp Hero was soon reopened as Montauk Point, followed by the Montauk Air Force Station. The base was said to have been closed for good in 1978 at the order of President Jimmy Carter. Montauk theorists, however, suggest that the work continued way below the old base, regardless of the fact that nothing at all was afoot on the surface, as the 1980s loomed on the horizon.

So, yes, there certainly was a military facility at Montauk—and it was a place that, at various points in time, was integral to the arsenal of the Air Force. That much is verified. Let's see what else can be verified.

It may come as a surprise to many to learn that the U.S. Navy of today does not deny that something may have happened at the Philadelphia Naval Yard in late 1943. They don't—you may already have guessed—endorse the tales of invisible sailors and a teleporting ship. Rather, the Navy believes that the legends were born out of legitimate programs that became sensationally distorted over time.

Because they are often contacted by people wanting to know about the Philadelphia Experiment, the U.S. Navy has a couple of user-friendly information sheets available about the matter, both of which outline what the Navy believes to have been the origins of the experiment. It's worth taking a look at both information sheets, as they provide data that suggest more than one answer to the riddle—from their perspective, at least. In its first document on the controversial affair, the Navy states the following:

> Allegedly, in the fall of 1943 a U.S. Navy destroyer was made invisible and teleported from Philadelphia, Pennsylvania, to Norfolk, Virginia, in an incident known as the Philadelphia Experiment. Records in the Archives Branch of the Naval History and Heritage Command have been repeatedly searched, but no documents have been located which confirm the event, or any interest by the Navy in attempting such an achievement.

> The ship involved in the experiment was supposedly the USS *Eldridge*. The Archives has reviewed the deck log and war diary from *Eldridge*'s commissioning on 27 August 1943 at the New York Navy Yard through December 1943. The following description of *Eldridge*'s activities are summarized from the ship's war diary. After commissioning, *Eldridge* remained in New York

and in the Long Island Sound until 16 September when it sailed to Bermuda. From 18 September, the ship was in the vicinity of Bermuda undergoing training and sea trials until 15 October when *Eldridge* left in a convoy for New York where the convoy entered on 18 October. *Eldridge* remained in New York harbor until 1 November when it was part of the escort for Convoy UGS-23 (New York Section). On 2 November the convoy entered Naval Operating Base, Norfolk. On 3 November, *Eldridge* and Convoy UGS-23 left for Casablanca where it arrived on 22 November. On 29 November, *Eldridge* left as one of escorts for Convoy GUS-22 and arrived with the convoy on 17 December at New York harbor. *Eldridge* remained in New York on availability training and in Block Island Sound until 31 December when it steamed to Norfolk with four other ships. During this time frame, *Eldridge* was never in Philadelphia.

A copy of *Eldridge*'s complete World War II action report and war diary coverage, including the remarks section of the 1943 deck log, is held by the Archives on microfilm, NRS-1978-26. The original file is held by the National Archives.

Supposedly, the crew of the civilian merchant ship SS *Andrew Furuseth* observed the arrival via teleportation of the *Eldridge* into the Norfolk area. The SS *Andrew Furuseth*'s movement report cards are in the Tenth Fleet records in the custody of the Modern Military Branch, National Archives and Records Administration, (8601 Adelphi Road, College Park, MD 20740-6001), which also has custody of the action reports, war diaries and deck logs of all World War II Navy ships, including *Eldridge*. The movement report cards list the merchant ship's ports of call, the dates of the visit, and convoy designation, if any. The movement report card shows that the SS *Andrew Furuseth* left Norfolk with Convoy UGS-15 on 16 August 1943 and arrived at Casablanca on 2 September. The ship left Casablanca on 19 September and arrived off Cape Henry on 4 October. The SS *Andrew Furuseth* left Norfolk with Convoy UGS-22 on 25 October and arrived at Oran on 12 November. The ship remained in the Mediterranean until it returned with Convoy GUS-25 to Hampton Roads on 17 January 1944. The Archives has a letter from Lieutenant Junior Grade William S. Dodge, USNR, (Ret.), the Master of *Andrew Furuseth* in 1943, categorically denying that he or his crew observed any unusual event while in Norfolk. *Eldridge* and *Andrew Furuseth* were not even in Norfolk at the same time.

The Office of Naval Research (ONR) has stated that the use of force fields to make a ship and her crew invisible does not conform to known physical laws. ONR also claims that Dr. Albert Einstein's Unified Field Theory was never completed. During 1943-1944, Einstein was a part-time consultant with the Navy's Bureau of Ordnance, undertaking theoretical research on explosives and explosions. There is no indication that Einstein was involved in research relevant to invisibility or to teleportation....

TIME TRAVEL: THE SCIENCE AND SCIENCE FICTION

The Philadelphia Experiment has also been called 'Project Rainbow.' A comprehensive search of the Archives has failed to identify records of a Project Rainbow relating to teleportation or making a ship disappear. In the 1940s, the code name RAINBOW was used to refer to the Rome-Berlin-Tokyo Axis. The RAINBOW plans were the war plans to defeat Italy, Germany and Japan. RAINBOW V, the plan in effect on 7 December 1941 when Japan attacked Pearl Harbor, was the plan the U.S. used to fight the Axis powers.

Some researchers have erroneously concluded that degaussing has a connection with making an object invisible. Degaussing is a process in which a system of electrical cables are installed around the circumference of ship's hull, running from bow to stern on both sides. A measured electrical current is passed through these cables to cancel out the ship's magnetic field. Degaussing equipment was installed in the hull of Navy ships and could be turned on whenever the ship was in waters that might contain magnetic mines, usually shallow waters in combat areas. It could be said that degaussing, correctly done, makes a ship 'invisible' to the sensors of magnetic mines, but the ship remains visible to the human eye, radar, and underwater listening devices.

After many years of searching, the staff of the Archives and independent researchers have not located any official documents that support the assertion that an invisibility or teleportation experiment involving a Navy ship occurred at Philadelphia or any other location.

Thus concludes one official press release from the U.S. Navy. The Navy expands on the story with its second report on the matter:

Over the years, the Navy has received innumerable queries about the so-called 'Philadelphia Experiment' or 'Project' and the alleged role of the Office of Naval Research (ONR) in it. The majority of these inquiries are directed to the Office of Naval Research or to the Fourth Naval District in Philadelphia. The frequency of these queries predictably intensifies each time the experiment is mentioned by the popular press, often in a science fiction book.

The genesis of the Philadelphia Experiment myth dates back to 1955 with the publication of *The Case for the UFO* by the late Morris K. Jessup. Some time after the publication of the book, Jessup received correspondence from a Carlos Miguel Allende, who gave his address as R.D. #1, Box 223, New Kensington, Pa. In his correspondence, Allende commented on Jessup's book and gave details of an alleged secret naval experiment conducted by the Navy in Philadelphia in 1943. During the experiment, according to Allende, a ship was rendered invisible and teleported to and from Norfolk in a few minutes, with some terrible after-effects for crew members. Supposedly, this incredible feat was accomplished by applying Einstein's "unified field" theory. Allende claimed that he had witnessed the experiment from another ship and that the incident was reported in a Philadelphia newspaper. The identity of the newspaper has never been established. Similarly, the identity of Allende is unknown, and no information exists on his present address.

"The genesis of the Philadelphia Experiment myth dates back to 1955 with the publication of The Case for UFOs by the late Morris K. Jessup."

In 1956 a copy of Jessup's book was mailed anonymously to ONR. The pages of the book were interspersed with hand-written comments which alleged a knowledge of UFO's, their means of motion, the culture and ethos of the beings occupying these UFO's, described in pseudo-scientific and incoherent terms.

Two officers, then assigned to ONR, took a personal interest in the book and showed it to Jessup. Jessup concluded that the writer of those comments on his book was the same person who had written him about the Philadelphia Experiment. These two officers personally had the book re-typed and arranged for the reprint, in typewritten form, of 25 copies. The officers and their personal belongings have left ONR many years ago, and ONR does not have a file copy of the annotated book.

Personnel at the Fourth Naval District believe that the questions surrounding the so-called "Philadelphia Experiment" arise from quite routine research which occurred during World War II at the Philadelphia Naval Shipyard. Until recently, it was believed that the foundation for the apocryphal stories arose from degaussing experiments which have the effect of making a ship undetectable or "invisible" to magnetic mines. Another likely genesis of the bizarre stories about levitation, teleportation and effects on human crew members might be attributed to experiments with the generating plant of a destroyer, the USS *Timmerman*. In the 1950s this ship was part of an experiment to test the effects of a small, high-frequency generator providing 1,000 hz instead of the standard 400 hz. The higher frequency generator produced corona discharges, and other well known phenomena associated with high frequency generators. None of the crew suffered effects from the experiment.

ONR has never conducted any investigations on invisibility, either in 1943 or at any other time. (ONR was established in 1946.) In view of present scientific knowledge, ONR scientists do not believe that such an experiment could be possible except in the realm of science fiction.

TIME TRAVEL: THE SCIENCE AND SCIENCE FICTION

Of course, that the Navy initially claimed there was nothing to the story but later provided two very different explanations has had some researchers of the case rolling their eyes and shaking their heads. And the story does not end there. There is verifiable data that the U.S. military, in the 1940s, was exploring the issue of invisibility. One such program was code-named "Yahootie." The plan was to create an airplane that could not be visually seen. The plan revolved around strategically placed lights and mirrors on the planes, which were designed to reflect the skies in which the plane was flying. Of course, this would not have amounted to literal invisibility, but it does show that some degree of invisibility was an issue on the minds of the military when the experiment was said to have occurred in Philadelphia in 1943. If nothing else, this brief aside is definitive food for thought.

Such is the complexity of this story that I have split it into two chapters. We have just addressed the matter of invisibility as it relates to the U.S. Navy. Sit back now and indulge yourself in the ways the bizarre experiment came to be associated with time travel. It all revolves around a place called Montauk.

CHAPTER 5

THE MAN WHO
TRAVELED TIME—OR DIDN`T

Although no official government documentation on the fringe topic of time travel has ever surfaced from the alleged secret archives of the now-closed Montauk Air Force Station of New York, it's a fact that the U.S. military has indeed taken a keen interest in incredible fringe subjects. We know this because the U.S. Air Force, under the terms of the Freedom of Information Act, released a startling document on the subject into the public domain. Its title: *The Teleportation Physics Study*. It was the work of Eric W. Davis of Warp Drive Metrics of Las Vegas, Nevada, a company secretly contracted by the Air Force Research Laboratory.

In his report for the Air Force, Davis wrote: "This study was tasked with the purpose of collecting information describing the teleportation of material objects, providing a description of teleportation as it occurs in physics, its theoretical and experimental status, and a projection of potential applications."

Interestingly, Davis appeared to imply knowledge of other research in this particular field. He wrote that it was known to him that "anomalous teleportation has been scientifically investigated and separately documented by the Department of Defense." Again, there is no smoking gun here, but what there is provides yet another example of intriguing research of the type said to be far more advanced at Montauk.

One of the truly strangest—many have said wholly outrageous—allegations that has been made within conspiracy-themed research circles is that Montauk has a connection to the United States' most famous of all monsters, Bigfoot. The claim is that top-secret research is afoot deep below the old base to create tulpa-style versions of Bigfoot. A tulpa is a creature conjured up in the imagination that can then be projected outwardly and given some degree of quasi-independent life in the real world. The website *Weird U.S.* notes that on one occasion, one of those attached to the secret experiments, a man named Duncan Cameron, envisioned in his mind "a large, angry, powerful Sasquatch-like" entity that "materialized at Montauk and began destroying the base in a rage. It utterly decimated the place, tanking the project and disconnecting it from the past. As soon as the equipment harnessing people's psychic power was destroyed, the beast disappeared." And, as we've seen, there's a time-travel angle to Bigfoot, too.

In a report for the U.S. Air Force, Eric W. Davis wrote that his company "was tasked with … collecting information describing the teleportation of material objects, providing a description of teleportation as it occurs in physics, … and a projection of potential applications."

Still on the matter of Montauk and mysterious creatures, Joe Nickell is a senior research fellow of the Committee for Skeptical Inquiry (CSI) and "Investigative Files" columnist for *Skeptical Inquirer*. A former stage magician, private investigator, and teacher, he is author of numerous books, including *Inquest on the Shroud of Turin* (1998), *Pen, Ink and Evidence* (2003), *Unsolved History* (2005), and *Adventures in Paranormal Investigation* (2007).

He notes in a column from spring 2012: "In July 2008, the carcass of a creature soon dubbed the 'Montauk Monster' allegedly washed ashore near Montauk, Long Island, New York. It sparked much speculation and controversy, with some suggesting it was a shell-less sea turtle, a dog or other canid, a sheep, or a rodent—or even a latex fake or possible mutation experiment from the nearby Plum Island Animal Disease Center."

The strange saga of the admittedly very weird beast was one that caught the attention of not just national news outlets but international media. This was hardly surprising, since the animal appeared to have a beak-like face, large claws, and a dog-like body. While the controversy rolled on for a long time and provoked deep rumors about what "the government" was doing, an answer to the riddle finally came, as Dr. Darren Naish noted at *Science Blogs*:

"Is the carcass that of a dog? Dogs have an inflated frontal region that gives them a pronounced bony brow or forehead, and in contrast the Montauk monster's

TIME TRAVEL: THE SCIENCE AND SCIENCE FICTION

Joe Nickell, a senior research fellow of the Committee for Skeptical Inquiry, pictured here at the CSICon 2018, wrote of strange doings in Montauk, New York, such as that of an alleged monster washed ashore in July 2008.

head seems smoothly convex. As many people have now noticed, there is a much better match: Raccoon *Procyon lotor*. It was the digits of the hands that gave this away for me: the Montauk carcass has very strange, elongate, almost human-like fingers with short claws."

Finally, there is the matter of time travel. For years, rumors have circulated to the effect that the alleged missing sailors from the USS *Eldridge* were not rendered invisible but were flung into the future, to our present. While this remains the most controversial aspect of the various claims about Montauk and, of course, has yet to be proved, it is also an issue that has some supportive data. Dr. David Lewis Anderson of the Anderson Institute, which is based in New Mexico, states that many years ago he worked on a highly classified program that was focused on time travel. The location was the U.S. Air Force's Flight Test Center at the California-based Edwards Air Force Base.

Perhaps one day we will know for sure the true story of Montauk. The revelations may prove to be amazing.

We are now about to get to the heart of the time travel of the story—all the way from Philadelphia in World War II to the modern era. Get ready to learn the story of the man who claimed to have traveled in time, after getting caught up in that legendary Philadelphia Experiment. His name: Al Bielek.

A writer for the paranormal investigative organization Orange County Ghosts and Legends states at its website: "Say what you will about the concept of time travel. Some of us may think the mere thought of it is pure nonsense. Regardless of what you think, I think we can all agree that it is awfully fun to think about. Take Al Bielek's story for example." Quoting a writer named Justin Andress, the Orange County Ghosts and Legends writer continues: "According to popular legend, in 1943, the U.S. Navy undertook secret experiments based out of the port of Philadelphia. These experiments were designed to put Einstein's unified field theory to practical use by making a naval ship invisible. While conspiracy theorists debate the existence of the Philadelphia Experiment, one alleged survivor of the scientific outing, Al Bielek, maintained that the Navy's purpose was entirely different. According to Bielek, the true purpose of the Philadelphia Experiment wasn't invisibility, it was time travel.

"In 1990, Bielek claimed that he spent time in two separate periods of the future only to return to the present and tell his story. And that was just the beginning of the fantastic revelations of this totally, completely, absolutely, one hundred percent not fake time traveler. As if someone would make that up anyway."

The staff at *Gaia* have taken a careful look at the story of Bielek. They say that "Tesla gave von Neumann a cryptic warning about a 'personnel problem' that might occur in their experiment, but he continued anyway and the Navy trained a crew specifically for the operation. Then on August 12, 1943, they ran a second test. After being shrouded in a 'green, ozone-laden haze' the ship purportedly disappeared for several hours, during which it traveled through time and then rematerialized. Upon its return, sailors were reported to be violently ill, some engulfed in flames, and others molecularly bonded with the ship. Bielek, however, said he and his brother, who were aboard the ship at the time, jumped off during the time warp and remained in 1983 on Montauk, Long Island at another secretive government facility also experimenting with time travel, known as the Montauk Project.

"Bielek gives an intriguing explanation for how his time travel was possible, relating to Tesla's Zero Time Generator. He said that Tesla's device was the key for the ship to return back to its original location. According to Bielek we live in a five-dimensional reality, with time being the fourth and fifth dimension. He said that every human is given a set of locks that lock them in a point of time from which

The Semi-Automatic Ground Environment (SAGE) radar facility at the Montauk Air Force Station, decommissioned since 1981. In the 1940s the base "was allegedly developing psychological warfare techniques that included studying the concept of time travel," according to the *Toronto Tribune*.

TIME TRAVEL: THE SCIENCE AND SCIENCE FICTION

they came, but that the experiment ruptured those time references, upon returning to Philadelphia."

Paul Fitzgerald, writing for the *Toronto Tribune*, said: "Al and Duncan's story started in 1943 on Montauk, Long Island. There, the Montauk Air Force Station was allegedly developing psychological warfare techniques that included studying the concept of time travel. The brothers were aboard the USS *Eldridge*, a top-secret naval ship that could supposedly turn itself invisible and teleport around the world. If that wasn't crazy enough already, Al and Duncan claim that they jumped off the ship late one evening, and instead of landing in the water, they traveled to the year 2137!

"When they arrived, they spent six full weeks lying in hospital beds recovering from intense radiation poisoning. There was no medical staff at the hospital, and they had no clue where the building was located. Apparently, the futuristic medical system that was treating their injuries used vibrations and light. Sounds nuts, right? Well …"

Writing at *Ice Pop*, Paige Steinman relates: "In 1988, Bielek and his family were watching a science fiction movie, *The Philadelphia Experiment*. It was said to be based on a true story, but what took place in the film seemed to be unfathomable for everyone except for Bielek. As he watched, he could not shake this unsettling feeling inside of him that he had seen this all before.

"The movie *The Philadelphia Experiment* depicted a government experiment back during the World War II era. According to the film, the U.S. Navy was determined to figure out a way to make it so that their ships were completely invisible to radar transmissions, and unable to be detected by other governments…. From the moment Bielek finished watching the film, he began to experience intense flashbacks. Bielek grew increasingly concerned about these remarkably vivid flashbacks, which felt like very clear memories. He was not sure if they were truly memories, or if he was going completely insane. He knew he needed to seek help to get to the bottom of whatever was happening."

A writer for the *Thrillist*, Dave Gonzales, offered words on this controversial matter, too: "After seeing *The Philadelphia Experiment* in 1988, 57-year-old Al Bielek couldn't shake the eerie feeling that he'd seen it somewhere before. Undergoing various forms of New Age therapies, Bielek was able to uncover repressed memories of having worked on the Montauk Project in the 1970s and '80s; he also ascertained that his memories had been locked away to keep the experiment secret. As his memories came flooding back, he learned that his name wasn't Al Bielek, after all; born Edward Cameron, he'd also worked on the Philadelphia Experiment with his brother, Duncan Cameron, when both men were in their mid-20s.

> *"Bielek claimed that, sometime in the 1940s, Nikola Tesla figured out how to make the USS Eldridge invisible and, in the process, opened up a time wormhole into the future that sucked in the ship."*

TIME TRAVEL: THE SCIENCE AND SCIENCE FICTION

"A few years later, Al Bielek presented his story at a Mutual UFO Network conference. The Philadelphia Experiment was real, he said, and he was the proof, having lived out the World War II section of the movie. Bielek claimed that, sometime in the 1940s, Nikola Tesla figured out how to make the USS *Eldridge* invisible and, in the process, opened up a time wormhole into the future that sucked in the ship. The Cameron brothers were on board, jumping off the vessel and landing at Montauk's Camp Hero—on August 12th, 1983."

The Bielek saga continues to provoke a great deal of controversy, and it's unlikely to go away anytime soon. The believers will continue to believe, and the doubters will continue to doubt. The only question that really counts is this one: what's more likely, that Bielek traveled through time or that he created a story based on an entertaining 1984 sci-fi/conspiracy movie titled *The Philadelphia Experiment?*

CHAPTER 6

TIME TRAVELERS DRESSED IN BLACK

One of the more intriguing rumors surrounding Area 51, a highly classified U.S. Air Force installation in southern Nevada, is the theory that it is the home base from which the legendary and ominous Men in Black operate—and, further, that the MIB may be time travelers rather than government agents or aliens in disguise. Make mention of the Men in Black to most people, and you will likely provoke images of actors Will Smith and Tommy Lee Jones, or perhaps the duo of Tessa Thompson and Chris Hemsworth. After all, the trilogy of *Men in Black* movies was phenomenally successful and brought the subject to a worldwide audience. Outside of ufology, most people assume that the Men in Black were the creations of Hollywood. This, however, is very wide of the mark: in reality, the movies were based on a short-lived comic book series that was created by Lowell Cunningham in 1990. Most important, the comic books were based on real-life encounters with the MIB dating back decades. As for those real MIB, they just might be skilled surfers of time.

In the movies, the characters portrayed by Jones and Smith are known as J and K. There is a good reason for that: they are the initials of the late John Keel, who wrote the acclaimed book *The Mothman Prophecies* and who spent a lot of time pursuing MIB encounters, particularly in the 1960s and 1970s. In that sense, the producers of the *Men in Black* movies and comic books were paying homage to Keel. Now let's get to the heart of the matter—namely, the real Men in Black, not those of Hollywood. Who are they? Where do they come from? What is their agenda? If there is one thing we can say for sure about the MIB, it's that they are the ultimate

CHRIS **HEMSWORTH**   TESSA **THOMPSON**

Chris Hemsworth and Tessa Thompson, pictured here, starred in the 2019 film *MIB: International*. Some suggest that real Men in Black are time travelers or biological robots sent from the future.

controllers: they threaten, intimidate, and terrify into silence those they visit. Let's see how the mystery all began.

It was in the early 1950s that a man named Albert Bender created a UFO research group called the International Flying Saucer Bureau (IFSB). The group was based in Bender's hometown of Bridgeport, Connecticut. Bender quickly became enthused by the UFO phenomenon when it kicked off in earnest in the summer of 1947 with Kenneth Arnold's acclaimed and now-legendary sighting of a squadron of UFOs over the Cascade Mountains. The world was changed, and so was Albert Bender.

As a result of the establishment of the IFSB, Albert Bender found himself inundated with letters, phone calls, and inquiries from people wanting information on the UFO enigma. Bender was pleased to oblige, and he created his very own newsletter, called *Space Review*. The publication was regularly filled with worldwide

Albert K. Bender, founder of the International Flying Saucer Bureau in the early 1950s, suddenly abandoned his beloved UFO work after a visit from three mysterious Men in Black.

accounts of UFO activity, alien encounters, and sightings of flying saucers. On the worldwide issue, it's worth noting that so popular was Bender's group and magazine that he found himself inundated with letters from all around the planet: communications poured in from the United Kingdom, Australia, countries in South America, and even a few from Russia.

Bender was on a definitive high: the little journal that he typed up from his attic room in the old house in which he lived was suddenly a major part of ufology. It's most curious, then, that in the latter part of 1953, Bender quickly shut down the International Flying Saucer Bureau, and he ceased the publication of *Space Review*. Many of Bender's followers suspected that something was wrong—very wrong. As it happens, their suspicions were correct.

When Albert Bender brought his UFO-themed work to a hasty end, a few close friends approached him to find out what was wrong. After all, right up until the time of his decision to quit, he was riding high and had a planet-wide following. It didn't get much better for Bender. So, his decision to walk away from all things saucer-shaped was a puzzle. One of those who wanted answers was Gray Barker. A resident of West Virginia and both a writer and a publisher who also had a deep interest in UFOs, Barker had subscribed to *Space Review* from its very first issue and developed a good friendship and working relationship with Bender, which provided an important reason for Barker to question Bender's decision.

At first, Albert Bender was reluctant to share with Gray Barker his reasons for backing away from the subject that had enthused him for so long, but he finally opened up. It turns out—as Barker wrote in his 1956 book on the Bender affair, *They Knew Too Much about Flying Saucers*—that Bender had been visited by a trio of men, all dressed in black, who warned him not only to distance himself from the subject but to completely drop it. As in forever. Somewhat of a nervous character at the best of times, Bender hardly needed telling once. Well, actually, he did need more than one telling: despite having the fear of God put in him, Bender at first assumed that what the Men in Black didn't know wouldn't hurt them. So, despite the initial threat, Bender chose to soldier on. It was a big, big mistake. When the MIB realized that Bender had not followed their orders, they turned up the heat to an almost unbearable level. Finally, Bender got the message.

TIME TRAVEL: THE SCIENCE AND SCIENCE FICTION

For Gray Barker, who recognized the dollar value of the story of his friend, this was, in a strange way, great news. The scenario of a mysterious group of men in black suits terrorizing a rising UFO researcher would make for a great book, thought Barker. Hence Barker's 1956 book. The problem was that although Bender somewhat reluctantly let Barker tell his story, Bender didn't tell him the whole story. Bender described the three men as being dressed in black suits and confirmed the threats, but that was about all he would say. As a result, Barker quite understandably assumed that the Men in Black were from the government. He suspected they were from the Federal Bureau of Investigation (FBI), the Central Intelligence Agency (CIA), or the Air Force. Barker even mused on the possibility that the three men represented all of those agencies. When Barker's book was published, it not only caught the attention of the UFO research community of the day, but it also, for the first time, brought the Men in Black to the attention of just about everyone involved in the UFO issue. A legend was born that continues to this day.

While Albert Bender didn't lie to Gray Barker, he most certainly did not share with him the full story. In fact, Bender had barely shared the bones of it. There was a good reason for that: the real story was far, far stranger than Barker could have imagined. Yes, Bender was visited by three men in black, but they were not of the kind that the U.S. government of the day might have dispatched. Rather, they fell into the domain of the supernatural, the paranormal, and the occult.

*In seconds, something terrifying happened: three
shadowy, ghostly, spectral beings started to
materialize through the walls of Bender's room.*

According to Bender, late one night, after toiling away at his old typewriter in his attic environment, he suddenly started to feel sick. He was overwhelmed by nausea, dizziness, and a sense that he might faint. Most curious of all, the room was filled with an odor of brimstone, or sulfur. The odor is associated with paranormal activity and has been for centuries. Bender lay down on the bed, fearful that he might crash to the floor otherwise. In seconds, something terrifying happened: three shadowy, ghostly, spectral beings started to materialize through the walls of Bender's room—yes, through the walls. They didn't need to knock on the door and wait for it to be opened. The silhouette-like trio then started to change: their shadowy forms became more and more substantial until finally they took on the appearance of regular men, but with several notable differences. Their eyes shone brightly, like pieces of silver reflecting the sun. Their skin was pale and sickly looking, and they were thin to the point of being almost cadaverous. They closely resembled the deadly vampires of old, which Bender loved to read about in his spare time.

Using telepathy rather than the spoken word, the three men warned Bender that now was the time for him to leave the UFO issue alone—leave it and never return. Or else. When Bender began to shake with fear, the Men in Black realized that they had gotten their message across, and they duly departed the way they had first arrived— through the walls. For days, Bender was in a state of fear that bordered on hysteria.

TIME TRAVEL: THE SCIENCE AND SCIENCE FICTION

Finally, though, he thought: why should I quit ufology? After all, I've done so much work, I don't want to stop now. So Bender didn't stop; he decided to take on the MIB and stand up to their threats. That was a very big mistake on the part of Bender.

In the days ahead, Bender saw the MIB again. On one occasion, late on a Saturday night, Bender was sitting in his local cinema, watching a new movie, when one of the Men in Black materialized in the corner of the cinema, his blazing eyes focused on the terrified Bender. The ufologist didn't hang around but fled the place. On the way home, Bender was plagued by the sounds of footsteps behind him, which seemed to be disembodied as no one was in sight. Eventually, the MIB returned to that old attic, which yet again caused Bender to fall seriously ill. Finally, after another week of terror and mayhem, Bender really was done. His time in ufology was over—for the most part.

The story of Albert Bender as told in the pages of Gray Barker's 1956 book *They Knew Too Much about Flying Saucers* was substantially correct, in that it told of how Bender was visited, threatened, and ultimately driven to leave ufology. Through no fault of his own, though, Barker was unaware of the supernatural aspects of the story and assumed that Bender had become a victim of the U.S. government. Finally, though, Bender came clean with Barker. Far from being disappointed, Barker was overjoyed, chiefly because he realized that he could spin the Bender saga into yet another book. This time, though, Barker let Bender write the story himself. Yes, despite being warned away from the flying saucer issue by the Men in Black, Bender somewhat reluctantly reentered the scene and wrote his very own story in *Flying Saucers and the Three Men*, which Gray Barker eagerly published in 1962. Many people in ufology were put off by the supernatural aspects of the story, and as a result, the book was relegated to the realm of obscurity for many years.

It's interesting to note that behind the scenes there was another group of men in black suits—and black fedoras—who were secretly following the Bender saga. It was none other than the FBI. In other words, although FBI agents were not literally Bender's MIB, the FBI certainly wanted to find out who they were. Thus, in a strange way, there were now two groups of MIB, distinctly different: the supernatural ones encountered by Bender and the MIB of government officialdom. The provisions of the Freedom of Information Act have shown that both Albert Bender and Gray Barker had files opened on them. Those same files make it clear that none other than the legendary FBI boss J. Edgar Hoover ordered one of his special agents to get hold of a copy of Gray Barker's *They Knew Too Much about Flying Saucers*.

After promoting his book, Bender again walked away from the UFO issue. This time, it was for good. Bender died in March 2016, at the age of 94, in California.

TIME TRAVEL: THE SCIENCE AND SCIENCE FICTION

In the years that followed Bender's encounters, the U.S. government would become determined to uncover the truth of the MIB. During the course of his research into the issue of the Men in Black, John Keel arranged a meeting with one Colonel George P. Freeman of the United States Air Force. Keel's interest was driven by the fact that Colonel Freeman had circulated a memo throughout the Air Force ordering everyone to be on guard for the Men in Black. Colonel Freeman's memo reads as follows:

Mysterious men dressed in Air Force uniforms or bearing impressive credentials from government agencies have been silencing UFO witnesses. We have checked a number of these cases, and these men are not connected to the Air Force in any way. We haven't been able to find out anything about these men. By posing as Air Force officers and government agents, they are committing a Federal offense. We would sure like to catch one—unfortunately the trail is always too cold by the time we hear about these cases, but we are still trying.

Only a few weeks after Colonel Freeman's memo was widely circulated, Lieutenant General Hewitt T. Wheless, also of the U.S. Air Force, wrote his own:

Information, not verifiable, has reached Hq USAF that persons claiming to represent the Air Force or other Defense establishments have contacted citizens who have sighted unidentified flying objects. In one reported case, an individual in civilian clothes, who represented himself as a member of NORAD, demanded and received photos belonging to a private citizen. In another, a person in an Air Force uniform approached local police and other citizens who had sighted a UFO, assembled them in a school room and told them that they did not see what they thought they saw and that they should not talk to anyone about the sighting. All military and civilian personnel and particularly information officers and UFO investigating officers who hear of such reports should immediately notify their local OSI offices.

It was this period of interest in the MIB on the part of the government that led to an extraordinary, and almost surreal, development.

Although the U.S. government had no real idea of who, or what, the real Men in Black were, there was a realization on the part of the government that the phenomenon of the MIB could be used to the advantage of the likes of the National Security Agency (NSA), the CIA, and military intelligence. It wasn't just the MIB who wanted UFO witnesses silenced; the government did, too. But the government was concerned about threatening UFO witnesses—American citizens, in other words—and being outed in the process.

The government therefore came up with an ingenious idea: they created a group within the heart of officialdom whose job it would be to keep people away from the really important parts of the UFO phenomenon. Threats, silencing, and intimidation were the orders of the day. But how was that successfully achieved? By having their secret agents dress and act like the real MIB that had terrorized Albert Bender and intruded on the life of Brad Steiger. In other words, they wore black suits, black sunglasses, and black fedoras, and they acted in a distinctly odd, emotionless fashion. The government really did not know (and probably still does not know) who or what the MIB really were, but that same government knew that they could exploit the phenomenon to their distinct advantage. Dressing as the MIB would offer the government an ingenious form of camouflage. It was a case of using fear to provoke the ultimate form of control.

As the 1960s became the 1970s, and then the 1980s and the 1990s, and now the twenty-first century, the issue of the existence of two different types of MIB—government agents and something supernatural—continued.

Now we come to the matter of the MIB and time travel. Joshua P. Warren is one observer who suggests that the Men in Black may be time travelers. His thoughts on this issue: "Why do the MIB dress like this? Why do we call them the Men in Black? Well, if a man puts on a black suit with a black hat and walks down the street in 1910, and you see that man, you would probably notice him. But, would you think there was anything too extraordinary, or too out-of-place about him? No, you probably would not. And if you saw a man walking down the street in 2010 wearing a black suit and a black hat, would you notice him? Probably, yes. But, would you think there was necessarily anything too extraordinary? No."

Intriguing, to be sure. Things don't end there, though.

Schwarz had a fascinating theory: maybe the MIB are historians, carefully studying our time and doing their utmost to learn of the human race of the twentieth and twenty-first centuries.

The late ufologist Berthold E. Schwarz, M.D., pondered on the idea that the Men in Black were biological robots sent by humans from their future to our present. But for what reason? Schwarz had a fascinating theory: maybe the MIB are historians, carefully studying our time and doing their utmost to learn of the human race of the twentieth and twenty-first centuries. As incredible as that may sound, there is good, solid data to show that the Men in Black are indeed some form of robot.

In his excellent catalogue *MIB Encounters*, Gareth Medway includes a story that suggests the Men in Black might be nothing less than biological robots. Let's take a look at this phenomenon in today's world of science. On January 13, 2020, the U.K.'s *Guardian* newspaper ran an article titled "Scientists Use Stem Cells from Frogs to Build First Living Robots." In part, the article states: "Researchers in the US have created the first living machines by assembling cells from African clawed frogs into tiny robots that move around under their own steam. One of the most successful creations has two stumpy legs that propel it along on its 'chest'. Another has a hole in the middle that researchers turned into a pouch so it could shimmy around with miniature payloads. 'These are entirely new lifeforms. They have never before existed on Earth,' said Michael Levin, the director of the Allen Discovery Center at Tufts University in Medford, Massachusetts. 'They are living, programmable organisms.'"

Do the Men in Black fall into the category of those "programmable organisms"? Let's have a look at what we know about the MIB. We'll begin with the story that Gareth Medway highlights. It reads as follows: "The Christiansen family of Wildwood, New Jersey, who had seen a UFO on 22 November 1966,

Some suggest that the Men in Black dress in timeless black suits so that, as they travel through time, their clothing won't attract attention.

were interviewed by 'the strangest looking man I've ever seen,' wearing a thin black coat, who introduced himself as 'Tiny' from the 'Missing Heirs Bureau.' He spoke in a high, 'tinny' voice, in clipped words and phrases like a computer, 'as if he were reciting everything from memory.' His black trousers were too short, and 'they could see a long thick green wire attached to the inside of his leg, it came up out of his socks and disappeared under his trousers.' John Keel commented that he had not heard of this feature in other MIB cases: 'Was Tiny wearing electric socks? Or was he a wired android operated by remote control?' He departed in a black 1963 Cadillac."

Now to the theme of this chapter. First and foremost, the MIB are nothing like Will Smith and Tommy Lee Jones. The MIB-themed movies are good fun. But by presenting the Men in Black as the employees of a secret organization that wipes out dangerous aliens, the assumption is that they are human. They are not. There are many weird issues surrounding the real MIB phenomena that strongly suggest the Men in Black are something else entirely. Over the years I have addressed more than a few different theories for what the MIB might be. The list includes time travelers, tulpas/thought-forms, demons, extraterrestrials, interdimensional creatures, and more. For now, let's stay focused on that biological robot angle.

None of this proves that the MIB are biological robots
from our future. But as an admittedly controversial
theory, it's not a bad one.

TIME TRAVEL: THE SCIENCE AND SCIENCE FICTION

The fact is that the MIB *do* seem to act in a strange and robotic fashion. There is the matter of their skin. More than a few people who have been confronted by the Men in Black close up have noticed something very strange, even eerie. In many cases, their faces appear somewhat plastic, not unlike a nightmarish, hideous mannequin or a creepy old doll come to life. Moving on, Gareth Medway notes that the MIB talk in a very odd way. He mentioned they seem to recite "everything from memory." I have several cases in my files mentioning such a fact. In one case, the witness said the MIB appeared to have no actual understanding of what they were saying. The wording, the witness suspected, was programmed. On top of that, there's the matter of the somewhat clumsy, jerky fashion of walking that the MIB display. This is not described in every case, but there are more than a few such reports. Finally, there is the angle of food. Or, rather, from the perspective of the Men in Black, a distinct lack of food. That's right: the MIB have a distinct aversion to food. They don't even seem to know what food is. Perhaps, if the MIB are biological robots, they get their sustenance in a distinctly different way from us. None of this proves that the MIB are biological robots from our future. But as an admittedly controversial theory, it's not a bad one.

CHAPTER 7

NOT QUITE DELOREANS, BUT CLOSE

As we have seen, there is a distinct possibility—some might say a *probability*—that the Men in Black are nothing less than time travelers. Keeping that carefully in mind, it's worth taking a look at a lesser-known aspect of the Men in Black controversy that we have not yet addressed: the modes of transport the MIB use when they are in our time frame. Interestingly, they are almost always seen driving old-style, 1950s-era cars. And the cars are always, appropriately, black. There's something else, too: witnesses to the MIB and their vehicles state that the cars have the ability to vanish from view, quite literally, or even take to the skies.

Not unlike the DeLorean that Michael J. Fox's character, Marty McFly, drives in the 1985 motion picture *Back to the Future*, the cars of the Men in Black are here one second and gone the next. That may very well be because those "Cars in Black" are exiting our time and entering another realm, and vice versa. In light of all that, I'll now demonstrate how easy it is to see how and why the MIB and those mysterious cars go together—and all in relation to time travel. Welcome to the world of the CIB—the Cars in Black.

Dr. Josef Allen Hynek, a well-renowned UFO authority who died in 1986, was provided with the details of a fantastically strange MIB encounter that occurred in a small Minnesota town in late 1975 and falls firmly into the CIB category. No UFO was seen on this particular occasion; the chief witness was harassed by the driver of a large, black Cadillac on a particular stretch of highway and nearly forced into an

The DeLorean that Michael J. Fox's character Marty McFly drives in the comedy *Back to the Future* is not the only vehicle reportedly used as a time-traveling device. Pictured here is a replica of the car used in the movie.

adjacent ditch. The irate man quickly righted his vehicle and headed off in hot pursuit only to see the black Cadillac *lift into the air and, quite literally, disappear in the blink of an eye.* See what I mean about the parallels between the world of Hollywood and the domain of real-world time travel?

UFO expert Jenny Randles, who penned an excellent book on the MIB called *The Truth behind Men in Black,* investigated a similar affair that took place in the United Kingdom in the summer of 1981. The witness, a man named Jim Wilson, had encountered something odd—but not too amazing—in the night sky and was, unfortunately, visited by two Men in Black. They were, of course, the same type of pale, skinny, sickly-looking MIB that terrified Albert Bender way back in the 1950s. The two men told Wilson, in stern and creepy fashion, that all he had seen in the skies above was a Soviet satellite orbiting Earth, specifically Cosmos 408. And, they added, he had better keep his mouth firmly shut about it.

The whole thing would have come to an end except for one thing: the MIB wouldn't leave. Night after night they would carefully position their Car in Black—a Jaguar of 1960s-era vintage—and stare intently toward Wilson's home with mad-dog eyes. Unsurprisingly, Wilson's anxiety levels began to rise. When things got to be too much, Wilson called the local police.

TIME TRAVEL: THE SCIENCE AND SCIENCE FICTION

As two uniformed officers approached the vehicle and prepared to knock on one of the windows, the black Jaguar melted away into nothingness.

After seeing the vehicle positioned outside Wilson's car several times and then managing to get a good look at the license plate, which the police were quickly able to confirm as being totally bogus, they carefully closed in, with the intention of speaking with the pair of MIB and finding out the nature of their game. Unfortunately, they never got the chance to do so.

As two uniformed officers approached the vehicle and prepared to knock on one of the windows, the black Jaguar melted away into nothingness. There was, not surprisingly, a deep reluctance on the part of the officers to prepare any kind of written report alluding to such an event in the station logbook. Just like the Men in Black themselves, even their vehicles are seemingly able to perform disappearing acts of a ghostlike nature.

Now I'll share with you the eerie story of Tracie Austin. Like me, she is a Brit who moved to the United States in the early 2000s. I've known Tracie since 1996 and have no doubts about her credibility. Here she tells her amazing story:

Tracie Austin recalls being approached, in the English countryside, by an old-style black Lincoln—an American car—that appeared to have no driver.

TIME TRAVEL: THE SCIENCE AND SCIENCE FICTION

I remember when me and you went to the *UFO Magazine* conference in September 1999, in Leeds [England]. I got the train there, you drove there, and then you drove me home, afterwards. Well, it wasn't long after that that this thing happened. I had moved up to Cheshire the week after I got back from the conference. While we were at the conference I had met Brigitte Barclay [a well-known alien abductee in the U.K.]. That was the first time I had met Brigitte. We just got talking and she happened to tell me of when she lived in the States [and] had Men in Black episodes. And how she was followed by black cars and black helicopters.

So, I went to the conference, talked to Brigitte, and then was the house move. A few days after, me and a friend went to the movies and a Pizza Hut—it was a Sunday night; I remember that. We were driving from Cheshire into Staffordshire, for about fifteen miles. And Cheshire is full of country roads, as you know. Farmers' fields, countryside; that kind of thing.

As we entered Staffordshire, I noticed there was an old, black Lincoln car coming towards us in the opposite direction. Although it was old, it looked new, but it was an old style, and it was totally out of place. And, coming from England, in Cheshire, that's just not the kind of car you see: an American Lincoln. That's what I thought: What's an American car doing here?

As the car approached us—coming the other way—I tried to see who the driver was, but I couldn't see anybody. I'm not saying the car was driving itself, but I just couldn't make out anyone. *No one.* I could see it had really shiny bumpers and there was no [license] plate, at all. And I noticed that with every car behind that Lincoln, it was like it was the leader of the pack. I could see the drivers of the cars behind were all kind of "switched off." It looked like an automatic mode. Like they were in a daze. It was very, very odd.

We got down the road and my friend said: "Did you see the way that vehicle was moving?"

I said, "No, what do you mean?"

He said: *"The wheels were going around, but they weren't touching the ground"* [italics mine].

It was just like it was hovering, but not high, or I would have known. I was too busy trying to see the driver. I just said: "Oh, my God." It was really odd that I had had that conversation with Brigitte only a week prior, and now, a week later, I was having my own Man in Black encounter. I don't believe in coincidence at all; I think everything is synchronicity.

TIME TRAVEL: THE SCIENCE AND SCIENCE FICTION

Here's another example of this matter of the Men in Black and their strange cars. It turns out that in 1967—when Mothman mania (an issue we shall return to) was at its height in Point Pleasant, West Virginia—matters relative to the MIB were going down in Defiance, Ohio, too. The mysterious story revolved around a man named Robert Easley. He was a UFO researcher and a good friend of Timothy Green Beckley, a longtime UFO sleuth and someone who, in the 1960s, managed to capture an MIB on film in Jersey City.

For Easley, a resident of Defiance, it all began in the early hours of July 11, 1967. He was woken up by the sound of his phone ringing. Easley quickly answered it, perhaps fearing it was someone with bad news. It was actually a woman who wanted to report a sighting of two UFOs over the town—a sighting that was still going on. Easley jumped into his car and sped off into the night. It turns out that Easley was shadowed all the way by a black Cadillac, very often the preferred mode of transport for the MIB. The car had no license plates, which is typical, too. As for the driver, who kept very close to Easley's vehicle, his upper body and head were silhouetted and dark. That was not good news. The Man in Black soon vanished, however, taking another road and leaving Easley worried and slightly paranoid. He would soon be back, though. On the night of July 15, the MIB returned, yet again in his black Cadillac, following Easley for a number of nail-biting miles.

Tim Beckley tells the story of what happened next: "When [Easley] pulled into his driveway, the unknown car sped off. Later that evening as he sat talking with his girlfriend on the front porch, the car came down the road and stopped right in front of the house, as soon as the topic of UFOs entered their conversation. Easley could feel the man looking at them. When they got off the subject the car left, but when they got back on it about an hour later, the same car came back again. It was as if the driver could hear what they were saying or read their minds."

Robert Easley, a UFO researcher, was shadowed on a 1967 research trip by a black Cadillac, often the preferred mode of transport for the potentially time-traveling Men in Black.

TIME TRAVEL: THE SCIENCE AND SCIENCE FICTION

CHAPTER 8

MORE ON THOSE MYSTERIOUS CARS

N ow let's turn to a well-respected author and investigator of paranormal phenomena, David Weatherly. David, too, has had his encounters with these eerie Cars in Black. He says: "I've had several weird encounters in the area around Area 51, on trips out there. One particular time—this was 2014—I was driving back from a conference in California. I hit Nevada pretty late and I realized that I would take a route that would take me past Rachel and Alamo. So, I would get the chance to stay there for a night. The only thing in the area is the Little A'Le'Inn [a UFO-themed bar]. I called ahead and checked with the guy who ran the little motel there in Alamo. Alamo is a good place to use as a base and sky-watch. He was fine and said to come along to the office and he would have a key there for me. It was a nice night, and with it being the desert, it was still warm. I rolled in—this little town is a speck, you know? There's nothing there; certainly nothing that stays open much past five o'clock.

> *"It was really odd: he didn't have a hat on, but he was wearing a black suit and a skinny, black tie—like an eighties tie."*

"I got in, and the motel was just one of those really old strip motels, probably from the sixties. I pulled off and parked my car and there were only a couple of other cars there, in the parking lot. I got my room key and came out of the office and closed the door. I had my luggage—a roller—and my laptop, so my hands were full. My room was only a couple doors down from the office. But, as I'm moving towards my room, I see a guy leaning against a car in the parking lot. It was really odd: he didn't have a hat on, but he was wearing a black suit and a skinny, black tie—like an eighties tie. He was staring at me; and he wasn't there a moment before. It really stood out; really strange. He looks at me with this stern look and he says: 'There's not going to be any coffee.' It was so surreal; I was trying to process all the pieces. A couple of funny things stand out in my mind. He had a military haircut, that black skinny tie, and his shoes were black dress shoes, but they were really shiny. Although he was trying to appear casual, he didn't seem relaxed. It was like it was forced. There was something very awkward about it. It was only a few steps to my door, so I put the key in and opened the door and I thought: I'm gonna see what this guy's deal is. I rolled my bag into the room and only had to take a step or two into the room, to toss my laptop onto the bed. And in those few seconds, when I turned around, and looked outside the door, he was gone.

"It's really curious because there's a handful of weird, little incidents like that out at Alamo. There's another one that was probably a month later. There's another hotel there that has cabins, right down the street from the other hotel I was staying at. And it has a restaurant in their building; pretty decent food. I was there with a friend; we were staying in one of the cabins for the weekend. We were going to do some sky-watching with night-vision. We were in the restaurant for lunch and there were probably only three other people seated—and the staff.

"These two guys come in: black suits. They have black sunglasses on, no hats. They walked in and they walked the whole restaurant, as if they were looking for

Lovelock Cave in Nevada was inhabited by humans for at least 4,000 years and is the site of suspected cryptozoological and paranormal activity.

TIME TRAVEL: THE SCIENCE AND SCIENCE FICTION

someone. But they also stood out, and they looked a little odd. They came back to the center of the restaurant, where we were sitting, and these two guys stopped in the center and one of them, I noticed, had his phone up and was taking pictures of the inside of the restaurant, which, of course, was kind of curious. The other guy says, to no one in particular, 'We don't want any fish in here.' And, they walked off.

"One other time, in 2015, we were out sky-watching near Area 51, late at night. And this van drove out around us, two or three times. Again, this was right around the cabins. There is a circular drive that goes around the cabins, and this van came around: old, a dark green in color. This guy pulled through and on about the third pass he put his window down. And he had rolled the passenger's side-window down, as we were on the passenger's side, too. You couldn't really see him very clearly, but he stopped and sort of made this proclamation and said: 'People looking in the sky often see things they shouldn't see.' And then he drove away. That was the last we ever saw of him."

Although Wikipedia states that the location of Nevada's Lovelock Cave is "restricted," it's actually very easy to find: it's situated south of the town of Lovelock in Pershing County. It's a sizeable, shadowy cave, around 150 feet in length and 35 feet in width, and has a great deal of history and controversy attached to it. Crypto-zoological controversy, one might say. Excavations that began in the early twentieth century revealed that the cave was inhabited for at least 4,000 years, and possibly even longer than that.

In 1911 a pair of miners, James Hart and David Pugh, hauled out from the cave tons of bat guano (shit, for the uninitiated). Their actions revealed something amazing: a large number of ancient artifacts that had been buried for an untold number of millennia. In the years and decades that followed, a massive number of incredibly old items were discovered, studied, and cataloged. Those items included weapons, baskets, food-storage containers, slings, and even duck decoys for use in hunting. Although archaeologists concluded that various tribes may have inhabited the caves over the years, certainly the most documented presence is that concerning the Paiute people, who flourished not just in Nevada but also Arizona, California, Utah, Oregon, and Idaho. They continue to flourish. Not only that, they have a most intriguing legend—one of monstrous proportions.

According to the Paiute, in times long, long gone, they waged war on a mysterious race of giant humanoids known as the Si-Te-Cah. They were massive, violent, rampaging humanoids that fed voraciously on human flesh. Reportedly, the last of the Si-Te-Cah in Nevada were wiped out in the very heart of Lovelock Cave. They were forced into its depths by the Paiute, who filled the cave with bushes and then set them alight. The man-monsters reportedly died from the effects of fire and smoke. It was the end of a reign of terror that had plagued the Paiute for eons.

While there are rumors of at least *some* remains of the Si-Te-Cah being found in Lovelock Cave in the early twentieth century, such a thing has not been fully con-

Some link the Paiute legend of the Si-Te-Cah, a race of violent giant humanoids, to the Men in Black and time travelers.

firmed. Granted, there are a lot of stories, but the skeletal remains of huge humanoids whose heights ranged from six and a half feet to *twelve* feet? Well, that depends on whom you ask. While there are no formally confirmed remains of such monstrous goliaths, stories certainly circulate to the effect that when the initial excavations began in 1912, the remains of a man who stood in excess of six feet, and who was covered in red hair, were found, apparently in mummified, preserved states. So the legend goes.

Of course, the reported physical appearance of the beasts—that they were humanoid, very tall, and covered in hair—has inevitably given rise to the possibility that the dangerous tribe the Paiute fought millennia ago, which they called Si-Te-Cah, were what we would call Bigfoot today. A battle to the death, deep in the heart of Lovelock Cave? That might well have happened. It's no wonder, then, that the saga of Lovelock Cave intrigues and fascinates monster hunters and cryptozoologists. It may also attract the attention of the Men in Black.

With this strange tale of Nevada's Lovelock Cave told, it's now time to return to David Weatherly, who opened up about his own odd experience at Lovelock Cave. David said, as we awaited a call from the late-night radio show *Coast to Coast AM*:

> This would have been May of 2016. [Researcher and writer] Dave Spinks and I were doing a series of investigations across Nevada and we had driven out to Virginia City and a couple of other areas. While we were doing these investigations, we spoke to a Paiute elder at Pyramid Lake [approximately 40 miles northeast of Reno] who told us the whole legend of the Paiute version of Bigfoot, the Si-Te-Cah.

> Ironically, when Dave and I were driving back across Nevada, heading east, I happened to pull off the interstate and realized when we pulled off the exit that we were in Lovelock. There was a little sign there with "Lovelock" on it. The Paiute elder had not mentioned the name of the town. When I saw it, I

was like: "Oh my gosh, Dave, this is Lovelock. This is where the cave is." I thought, *We've got to find it.* So, we drove into the downtown area. We went to the local library and asked for directions. It's almost twenty miles out of town and you literally turn off the pavement after about a mile, and you're on dirt road. The librarian said there was an archaeological sign there to look out for. But, it's in the middle of nowhere. She said: "Nobody ever goes out there." This was in the middle of the week; it was a Thursday. We were pretty sure there wasn't going to be anybody out there. So, we start this drive.

It was road-trip time—a road trip that proceeded to get infinitely odd. And very quickly, too. Back to David:

We get maybe a third of the way out there on these dirt roads: there's nothing, no houses, no buildings, no nothing. And, I'm cruising along, I was driving. And I habitually check my rear-view mirror. All of a sudden there's a car behind us. It was really strange because there were no other turn-offs or anything. Dave was shocked, too. We were puzzling over where in the world did this car come from. It got stranger because this car maintained a regular distance between us. I just felt odd about it. It was a regular car, like a small Toyota. So, I hit the gas on the jeep, and I got up to probably 50, 60, 70 miles per hour on a dirt road. This car maintained the same distance behind us. He sped up and kept the same spacing. Which in itself was really odd. I'm kicking up a ton of dust, you know? So, then I slowed way, way down. And, again, he maintained the same distance.

Now, the road out there is a pretty straight stretch. But it finally comes to a point where there's a big bend in the road that hooks to the left. When I turned, the road started to slope up—up towards the cave. When I turned that bend and started to climb, I looked back and this car was gone. It had just disappeared. We get out to where the cave is, and there is a small parking lot and sign from the archaeological dig they did there many years ago. And

On a road trip along back roads through Lovelock, Nevada, paranormal phenomena investigators David Weatherly and Dave Spinks were followed by a mysterious Toyota.

TIME TRAVEL: THE SCIENCE AND SCIENCE FICTION

there's a little bathroom, but that's about it. So, we pulled into a parking space and both of us jump out.

We've got binoculars and we're looking for this freaking car. Well, we spotted it. Right before that bend, it had pulled off the road into the brush. It was sitting on the side of the road. Now it's all the more puzzling. I'm thinking, *What in the heck is this person doing?* So, we think, we're not on a schedule or anything, so we decided we're going to stay in the parking lot and see what this guy does. It was probably 20 minutes to a half hour before finally this car comes creeping up. The car pulls into the parking lot. There are only a handful of spaces in this lot. He parks it almost as far as he can from where my jeep is. And out jumps this young Asian guy. He has a hat on and he kind of jumps out of the car and he looks around with an odd gesture, and he says: "Aren't there supposed to be caves or something up here?"

Dave, who is closer to him, just points as it's pretty obvious from the sign. Dave says: "Right there." He pulls his hat down, rushes around to the trunk of his car, and pops the trunk open. This was in the Nevada heat; this kid digs around the trunk and gets a coat—a flannel coat. The whole time he's trying to hide his face. He doesn't want us to see his face. We were trying to take pictures of him—very covertly. Then, he goes over to the billboard; there's this little artistic drawing and it says what life was like for the natives. It's a painting of a few native people, a handful of little comments about what they did, how they lived. He stands there for probably another twenty minutes just staring at this thing. Then he shouts over to us: "I guess I'll go up there." And he takes off.

That was hardly the end of the matter, though, as David relates:

Now, the trail does a big loop, if you want to take the long trail to get there. It's about a quarter of a mile. So, he takes off up this loop. We watch him, and he gets around this rise and it kind of goes over some rocks and then he's gone, because the trail goes down the other side. Then it comes up and approaches the cave from behind. So we waited, again, for another half hour, because we wanted to do some filming for a project we were doing. Then, we decided to hit the trail and we're filing as we walk up the trail. Dave had the camera and I'm talking a bit about the caves and the legend. We get up to the rise and the trail hooks around to the left and we're still walking and all of a sudden this kid is behind us. It was the strangest damned thing.

This kid had actually snuck off the trail and was hiding behind some rocks, waiting for us to go by. So, we just can't believe this. So, we stopped and we're glancing back at us. And every time we stopped, he stopped. He'll bend down, pick up a rock; look at a weed, or something like that. And, again, it's very forced. So, we do about two-thirds of the trail and there's an outcropping of rocks to the left. And to the right there's a lookout point. We climb this outcrop and we're sitting out and we just waited. And this kid, he hemmed and hawed, trying to delay and looking at things and, finally, he goes past us. He goes over to the lookout point and he kept glancing over at us.

Then things got very surreal, as David makes abundantly clear:

> **Just like the CIBs, there is good evidence to show that the Bigfoot creatures are nothing less than time travelers, too.**

We're taking pictures on our cell phones and he glances at us at one point and sees us with the cell phones. Then, he reaches into his pocket and pulls out a cell phone. And this is what he's doing: looking at us as if he's trying to understand how to use it. He's stealing glances at us, like to make sure he's doing it right. Very weird. He stands there at this lookout point for a while, pretending to take pictures. Then, finally, he continues on. We let him go, and we wait and we wait. Probably another 20 minutes, maybe half an hour. Then, we slowly make our way up to the rest of the trail, where the cave is, figuring that by now surely this kid has gone, as the cave is small. As we get to the entrance of the cave, he pops out like some mad cuckoo. He rushes past us; he's in a sudden hurry to get out of there, and he runs down the trail. We spent probably an hour and a half or two hours in the cave, because we wanted to check out a lot of different things related to the reports. And we did some filming, too. Took a lot of photographs inside and out. We took our time.

Well, when we finally left the cave and we took that corner, where you can see the parking lot, there's that kid sitting in his car. And as soon as we turn the corner and start to trek down, he hits the gas and goes flying out of the parking lot. As if, again, he waited to see exactly how long we were going to be there and what we were going to do. It's funny: some elements reminded me of the John Keel story of the Oriental Man in Black who stole an ink pen [a story told in Keel's *The Mothman Prophecies*]. And, I should add, this car he was driving: no license plates at all.

We make our way down to the jeep and head back to town. And when we get to about where the pavement starts, going into Lovelock, another vehicle comes toward us—pulls over very dramatically and suddenly onto the side of the road. Now, this is an all-black vehicle, with blacked-out windows. By now, we're not paranoid, but we are thinking there's something really odd going on here. So, I'm sitting at a stop sign, watching this car do a very sudden U-turn to come and get behind us. I turned at the stop sign, and the entrance to the freeway is right there, so I jumped on the interstate. This car was coming up behind us, basically trying to catch up; he got caught up at the stop sign as there was some other traffic. But, I left him back somewhere.

At the least, this is all weird synchronicities. The final anecdote is that from the time we were there at Lovelock Cave and had all these weird incidents, when me and Dave now speak on the phone, we get weird interference. Noises, clicking sounds. It doesn't happen when he speaks to other people; it doesn't happen when I speak to other people. But it does if we bring something up about the cave.

There's a very good reason why I tie in the story of David and those Cars in Black and Bigfoot: just like the CIBs, there is good evidence to show that the Bigfoot creatures are nothing less than time travelers, too. Prepare to have your mind blown when we discuss this further in the chapter "Bigfoot, Wormholes, and Time Portals"!

CHAPTER 9

PROPHETIC DREAMS OF A NIGHTMARISH FUTURE

In this chapter we address a wholly alternative way of heading into the future—a way that we have no real control over and that doesn't require the use of a time machine to see what is to come. It's the matter of prophetic dreams—or, in some cases, absolute nightmares. Is it possible that while we are deep in our sleep states, we can head off into years to come? Maybe even decades and centuries? Based on what you are about read, the answer is a definitive "yes"! Before we get to the story, however, let's take a look at what prophetic dreams really are and how they can thrust us into future times. A writer for *Medium.com* comes right to the heart of the matter: "Prophetic dreams are dreams where you have an experience that has yet to happen in our realm of existence *but later happens*, where you will experience déjà vu from already having seen the result of that experience" (emphasis mine). In other words, prophetic dreams allow someone to see the future without the need for a time machine.

Dale M. Kushner at *Psychology Today* states the following, something that expands the understanding of the phenomenon: "In September of 1913, Carl Jung, the great pioneer of depth psychology, was on a train in his homeland of Switzerland when he experienced a waking vision. Gazing out the window at the countryside, he saw Europe inundated by a devastating flood. The vision shocked and disturbed him. Two weeks later, on the same journey, the vision reoccurred. This time an inner voice told him: 'Look at it well; it is wholly real and it will be so. You cannot doubt it.'"

In his 1962 memoir *Memories, Dreams, Reflections*, Jung said of this "waking vision": "I was suddenly seized by an overpowering vision: I saw a monstrous flood covering all the northern and low-lying lands between the North Sea and the Alps. When it came up to Switzerland I saw that the mountains grew higher and higher to protect our country. I realized that a frightful catastrophe was in progress. I saw the mighty yellow waves, the floating rubble of civilization, and the drowned bodies of uncounted thousands. Then the whole sea turned to blood."

The famous Swiss psychiatrist Carl Jung wrote of a "waking vision" that might have been a glimpse into the future.

Back to *Psychology Today*: "The following spring of 1914, he had three catastrophic dreams in which he saw Europe was deluged by ice, the vegetation was gone, and the land deserted by humans. Despite his awareness that the situation in Europe was 'darkening,' he interpreted these dreams personally and feared he was going mad. However, by August of that year, his dreams and visions were affirmed: World War I had broken out."

> *"Prophetic dreams are linked to major disasters, wars, assassinations, accidents, lottery numbers or even winning horse races."*

The people behind the web page *Prophetic Dreams* provide us with more insightful data: "Scientists rationalize that dreams that seemingly come true may actually be a suggestion or guide for you to follow, thus making it seem as if the dream is coming true. You are willing it to be true. Another argument is that you may have a tendency to self-edit your dream to match the outcome. Because dreams are so easily forgotten, your memory of the dream may not be accurate. Yet another theory is that your dreaming mind is able to piece together bits of information faster that your conscious mind. Your mind is able to see what will happen based on information that it has already collected. Still, there are believers that dreams can indeed predict or foretell the future. Prophetic dreams are linked to major disasters, wars, assassinations, accidents, lottery numbers or even winning horse races."

Prophetic dreams can be absolutely terrifying, too. As you are about to learn.

Midway through the summer of 2017, something unforeseen and very unsettling occurred. In the second week of August three people contacted me with eerily similar stories of atomic Armageddon in the near future. Not only that, in their nightmares the U.S. president (not Trump or Biden) is assassinated by North Korean agents, secretly working on behalf of Russia. When the U.S. government finds out the truth, North Korea is, in essence, turned into radioactive dust. But it doesn't end there: Russia is soon dragged in, and on the third day, nuclear weapons are used in Europe. The conflict grows and grows. In less than a week, an all-out nuclear conflict erupts. In less than a couple of hours, Europe, China, Russia, and the United States are destroyed. Billions are dead, all as a result of that assassination of the U.S. president. To receive numerous such eerily similar accounts was, I have to admit, chilling.

Then, on August 8, I received a Facebook message from a guy named Kenny who'd had a horrific dream of nuclear war two nights earlier. In Kenny's dream, the U.S. president was shot to death by a foreign agent, something that led to a nuclear war. Kenny lives in San Bernardino, California, and woke up suddenly in the dead of night in a state of terror. As Kenny explained, in his dream he was sitting in the living room of a house in a small town outside Lubbock, Texas. Kenny had no idea of the name of the town, only that he knew it was near Lubbock—a place he has never visited. In the dream, Kenny heard a sudden and deep rumbling sound that seemed to be coming from somewhere far away. He walked to a screen door, puzzled, and peered outside. To his horror, Kenny could see, way off in the distance, the one thing none of us ever want to see: a huge, nuclear mushroom cloud looming large and ominous on the horizon.

Kenny continued that in his dream, he was rooted to the spot, his legs shaking and his heart pounding. He could only stand and stare as the huge radioactive cloud extended to a height of what was clearly miles. The entire sky turned black, and suddenly a huge wave of flame and smoke, hundreds of feet high, raced across the entire landscape, completely obliterating everything in its path. In seconds, there was another explosion, again far off in the distance, but from the opposite direction. Nuclear war had begun. That was when Kenny woke up, thankful that it was all a dream but disturbed by the fact that, as Kenny told me, the dream seemed like something far more than just a regular dream. Kenny felt he had seen something that was yet to come: a glimpse of the near future.

Further dreams of nuclear nightmares came my way. Kimberly J. emailed me on August 10 to share a story of disturbing proportions. She lives in Chicago, Illinois, and had a dream somewhat similar to that of Kenny, but a few days later than his, on August 9. The scenario was almost identical: a gigantic explosion destroyed her home city, killing millions and vaporizing everything for miles. A huge mushroom cloud was hanging where, only seconds earlier, there had been a bustling city of close to 3 million people. In this case, however, there was more: amid the carnage and the chaos, a large "birdman," as Kimberly described it, hovered over the massive cloud, "watching the end of us." The reference to the "birdman" provokes imagery of the Mothman of Point Pleasant, West Virginia, which is connected to death, even to the extent of having its very own "death curse" and prophecies attached to it. Sightings of the Mothman, indeed, culminated in the tragic collapse of the city's Silver Bridge and led to the drownings of dozens of locals.

It just so happens that two months earlier, M. J. Banias wrote an article at the *Mysterious Universe* website titled "Chicago's Current Mothman Flap 'A Warning,' Says Expert." In his article, dated June 7, 2017, Banias described a then-recent wave of Mothman-type sightings in and around Chicago. The article quoted researcher Lon Strickler, who

Sightings of the Mothman or a Mothman-like creature have been linked to prophecies of future events.

looked into these particular cases, leading to the publication of his book on the Mothman–Chicago wave titled *Mothman Dynasty: Chicago's Winged Humanoids*. Strickler said:

> There are many opinions as to why these sightings are occurring, including a general feeling that unfortunate events may be in the city's future.... At this point, I feel that this being may be attempting to distinguish a connection between locales within the city and future events. The witnesses have been very steadfast with what they have seen, and refuse to embellish on their initial descriptions. Each witness has had a feeling of dread and foreboding, which I believe translates into a warning of some type.

Then, on August 12, I received yet *another* Facebook message of a similar nature. This one was from Jacob, an American who is now a resident of Mulhouse, France (oddly enough, it's a city I spent a lot of time in during my teenage years). In Jacob's dream, an emergency broadcast message appeared on his TV screen, warning people to take cover: the nukes were flying. And that was it: just a few brief seconds of mayhem in the dream state. But it was still an undeniably nightmarish night for Jacob.

Of course, all of this could have been as a result of the genuine, growing tensions between North Korea and the United States during the Trump administration. In fact, I'm sure that has a great deal to do with it. On August 9, the United Kingdom's *Independent* newspaper ran an article on the North Korea issue that stated, in part: "While it's unclear if North Korea can successfully target US cities like Denver *and Chicago* [italics mine] with a nuclear ICBM, it's similarly unknown if US defense systems can strike it down—adding to American anxieties." The issue of Chicago being a possible target had been mentioned in multiple news outlets. Such stories almost certainly would have been worrying to Kimberly J., who lives in the heart of the city.

> *I wondered: Should we be concerned that three people,*
> *inside one week, had nightmares about nuclear war?*

I wondered: Should we be concerned that three people, inside one week, had nightmares about nuclear war? And what about that connection to presidential assassinations? I decided I wouldn't be surprised if hundreds of people all around the world have had such dreams. Maybe even more. After all, the climate was hardly a stable one. It was, and still is, a time of intense anxiety. And Kimberly's sighting in her dream of a Mothman-type creature—which had been seen around Chicago in recent months and was linked to a possible looming disaster in the area—would likely have led some to believe there was more to all of this than just bad dreams. There is also the fact that I got a cluster of such reports across a very short period of time, which was not exactly something I considered to be good news.

Then there was this, from Stephen Polak: "As a Chicago resident myself who has recently had a dream of being consumed [by] an enormous wall of fire, I find all of this rather disquieting." Five days later the news got worse. Chris O'Brien, a well-known author of many books, including *Stalking the Herd* and *Secrets of the Mysterious Valley*, contacted me with a story to tell. It was not a good story.

Chris said: "Back in 2005 Grandfather Martin Gashweseoma, for many decades the 'Fire Clan Prophecy Tablet' holder, spent a week with Naia and I at our home in Sedona, AZ. We had met him 10 years prior and we had become friendly with the then 83-year-old Traditional Elder. During one conversation about the predicted 'End of the Fourth World,' I asked him how the dreaded 'War of the Gourd of Ashes' would end. (In 1989, Martin announced the start of the final conflict would begin within the year, and it did with 'Desert Storm.') He said that North Korea would send fiery birds high in the sky to the US. I pressed him for further details suggesting maybe he meant China, and he said, 'No, Korea will be behind this attack, possibly w/ the help (or at the behest) of China.' At the time Korea had no functioning nuclear weapons program and no ICBMs. As we all know, this has changed…. Just thought I'd mention this!"

TIME TRAVEL: THE SCIENCE AND SCIENCE FICTION

Several correspondents have experienced possible visions of a near future involving nuclear war.

The next development came on August 17, from Jason M.: "I also have had a very powerful, lucid dream—in which Orlando, Florida (which is about two hours from me) was hit by a massive blast followed by a tremendous fireball and mushroom cloud. The dream felt incredibly real, and I was even able to interact realistically with those around me and see and feel their fear. It was truly horrific. As someone who has had dreams that have come to pass in great detail, I took serious note of the dream, and I spoke with my wife about it moments after I woke up. The dream was also notable to me because I recall specific details and dates about the attack, because in the dream I was reading a news article about a geopolitical crisis that was rapidly spiraling out of control. It left me with an overwhelming sense.

"Here is the caveat. In my dream, I knew without doubt that the attack came from Russia, and those around me expressed a similar sentiment. In the news article I had been reading, it discussed how a crisis in the Ukraine had provoked a Russian invasion and a NATO response. I had this dream in early March 2014, just days before the Russian invasion of Crimea. To say that development freaked me out is a tremendous understatement. In the dream, the attack took place in September of 2015. I have no doubt of that.

"I was very apprehensive throughout 2015 and kept a close eye on news events from Ukraine. While I was relieved that nothing happened, I knew that the dream was more than just fear playing out in my unconscious mind. Since that time, I have revisited the dream many times in deep meditative states to see what more I could learn from it.

"I do think it is worth cautioning that such dreams could have many potential causes other than predicting actual nuclear annihilation. At least one can hope."

"What is my point here? I know that such a dream can be horrible and terrifying. It could even conceivably foreshadow coming events. However, I suspect something else is at play. I admit that I cannot be certain, but my conclusion about my own dream is that my unconscious mind was tapping into collective awareness and fear that was about to engulf the world about that particular crisis. I think I was seeing a potential outcome that was informed by the fear and imagination of millions around the world (or would be, since the crisis had not yet happened when I had the dream), not one that was fated to happen. I suppose there is no way to know but to wait and see. But I do think it is worth cautioning that such dreams could have many potential causes other than predicting actual nuclear annihilation. At least one can hope."

On August 17, a writer called Red Pill Junkie waded in with his own nightmare: "I don't live in Chicago but I had a similar dream on the night of August 6th (I remember the date well because I commented upon it with a friend on FB). I was with a group of people I didn't know and I looked outside and watched a ginormous mushroom cloud. It took a while for the thing to register but once everybody realized what we were seeing we all panicked and fled in search of refuge. I remember some tried to hide underground, but the blast buried them all alive. Then I woke up. I think this is the first time I've had a dream about nuclear Armageddon in a very long time."

On August 18, J. Griffin, commenting on the Chicago issue, said: "I'm in Chicago right now on business—the last street to my first destination was 'Nuclear Drive.' Go figure." Jacqueline Bradley sent me the following on August 21—the details of a dream that seemed to involve small, tactical nuclear weapons in a forthcoming confrontation: "A few days ago I had a dream that several nuclear events occurred—in my dream I remember the term 'thermonuclear.' There were several of these events popping up (appeared to be everywhere and small versions of what we would ordinarily be aware of). No one seemed to be very perturbed by these and people were just walking around, occasionally looking around and watching these. I was aware that if you were caught up in one and died it killed off your soul or spirit too. All this was happening in broad daylight on sunny days. The dream ended where I was in some kind of alley with an old-fashioned dustbin nearby. Suddenly I found myself 'sinking' or evaporating and woke up. I wasn't scared by the dream, just puzzled. I too connected it with the tensions in N Korea. I've also been watching *Twin Peaks* and connected it with that, but not sure why."

One day later, Jill S. Pingleton wrote: "As a paranormal investigator and student of metaphysics, I, like many, am concerned about the prophetic potential of so many having these dreams/visions. However as a former MUFON Chief Investigator, I'm wondering if the people reporting these dreams and associations are Contactees?" By "contactees," she is referring to people in contact with extraterrestrial entities on a regular basis.

TIME TRAVEL: THE SCIENCE AND SCIENCE FICTION

Jill continued: "My point is that Contactees frequently recount stories of viewing scenes of mass destruction placed in their mind's eye during encounters with ETs. I don't know if they are being given glimpses of the future or only a possible timeline unless events can be changed. Like a wake-up call to Contactees to get involved and speak out for the sake of humanity. Perhaps that's also the mission of the Mothman. I wonder if any of these dreams/visions were preceded by an abduction event or if it's part of an ongoing 'download' that so many Contactees experience. I think much can be learned from studying the Experiencers/Witnesses. So many questions!"

Over at *Red Dirt Report*, Andrew W. Griffin commented on an August 2017 article I wrote on the matter of Mothman and nuclear nightmares. In a feature of his own titled "Riders on the Storm (Strange Days Have Tracked Us Down)," he said: "Clearly we are entering very troubled waters. And it seems that the collective unconscious of humanity is clueing in that we are entering a perilous period in our history. So, when I saw Nick Redfern's new post at *Mysterious Universe*—'Mothman and Nuclear Nightmares'—I took pause, as he notes that 'in the last week, three people have contacted me with eerily similar stories' involving nuclear apocalypse."

Interestingly, Andrew added that in relation to Kenny's dream of a nuclear bomb exploding near Lubbock, Texas, it was "not unlike my own dream that I wrote about on Jan. 26, 2017, which involved nuclear detonations near Joplin, Missouri." He elaborated: "We were in a car in the vicinity of Joplin, Missouri—something I noted in my mind in that it is on that nexus of high weirdness 37 degrees north and 94 degrees west (which I recently addressed here)—and nuclear explosions, followed by menacing mushroom clouds, are going off at various intervals.... And yet as the nuclear blasts send radioactive debris through the town and infecting everything in its path, I seem to be the only one alarmed by what is happening around us. The whole experience has the feeling of a guided tour through a park or historic site."

Then there was a Facebook message from Andy Tomlinson of the city of Manchester, England. Back in early June of 2017, Andy had a dream of being in a deserted London. The city was not destroyed or in flames. It was, said Andy, "like they had all been evacuated," which is an interesting phrase to use. And the city was deserted except for two things: one was the sight of "a massive big black bird over [the Houses of] Parliament." Then, as Andy walked the streets, trying to figure out what had happened, he had that feeling of someone watching him. He turned around to see a man in a black trench coat right behind him. The man was pale and gaunt and, as Andy worded it, "had a funny smile." Andy's description sounds very much like a certain sinister character in the saga of the Mothman—one Indrid Cold, who, as we shall see in a later chapter, might have been a time traveler. Andy then woke up with his heart pounding, relieved that it had just been a dream. Or, was it something more than just a dream?

TIME TRAVEL: THE SCIENCE AND SCIENCE FICTION

Andy Tomlinson of Manchester, England, described a dream of "a massive big black bird" flying over the Houses of Parliament and an encounter there with a pale man wearing a black trench coat.

What was without doubt one of the most chilling stories came from Anna Jordan. On August 25, she sent me this:

"Hi Nick. So, this dream was one I had about 25 years ago or so. Here's the dream: I was standing in the living room of the apartment I lived in at the time. I had three small kids on the floor around me playing and *Sesame Street* was on the TV. (At the time, I only had two children, but later had a third.) So, PBS is on with *Sesame Street*, but there's a break-in on the channel like when they have breaking news. It just showed like a PBS symbol and a countdown and then this man, who I knew was in the White House, came on. He was sitting at a desk, very solemn. I swear to you, it was Mike Pence. I have waited, studying White House faces for all these years waiting to see this face. The white hair, the face, the voice … it's him, I'm sure. He just looks at the camera and says that he was very sorry, it was too late, and wished the world good luck. I don't know how, but I knew he was talking about a nuclear war. In the dream, I felt my heart drop. Then I woke up. That's about it."

Less than 24 hours later, Elaine Clayton sent me the following message at Facebook:

"Over the decades, but more so recently, I've had dreams of hologram writings and ships in the sky. I believe some of these dreams were actually astral travels (being scanned in a space ship, etc.) and I used to be woken up by—and this is the best way to describe it—robotic forms with hologram-like presence, brightly colored and with personality. I asked to stop seeing those when I could not tell who they were although they seemed benevolent. Those were visions, they happened with my eyes open sitting up. But most dreams are about spaceships of magnitude often geometrically fascinating. And several have shown me that in the distance that I perceive as 'in the West' [are] atomic bombs going off although they're only more like dirty bombs, not fully atomic.

"The last one went like this: I was standing looking up at a silently moving, ethereal looking space ship. It was huge and had smoke streaks coming off it toward earth. It was colorful and its structure appear[ed] smoky, multidimensional. There was a peaceful feeling more than military or fearful. Although again it was extremely dominant and intelligent energy. I then turned to look at the landscape behind me and saw all at once about 5 or 6 bombs being dropped, immediately exploding in small mushroom clouds of fire. I registered my sense of where they landed—to the north east of Manhattan, I thought, and knew they missed their target.

"But in the dream I worried about my sister, knowing she was there. But then when I woke up I figured it out—my sister lives in Colorado Springs [Colorado] and very near NORAD. I believe whoever dropped the bombs dropped them from a satellite with the intention to take out NORAD. But they missed. I later learned that Kim Jong Un has a map showing his plan to bomb NORAD but his map is not smart and he'd actually be bombing Louisiana. I may post this on my own site. I am so glad you study all these things."

As August came to its close, I was contacted by an old English friend, Sally, whom I had not seen in years. She suggested that I should read *Warday*, a 1984 novel written by Whitley Strieber and James Kunetka. I had actually read the book years ago, and it made for grim and disturbing reading. It's an excellently written book that tells the story of a limited nuclear attack on the United States that nonetheless kills more than 60 million people, some from the initial atomic blasts and others as a result of famine and starvation, radiation, and a wave of out-of-control influenza. Strieber and Kunetka skillfully tell a story that could, one day, become all too real. In *Warday*, the United States is a shell of its former self, rampant with chaos, death, and destruction. *Warday* makes it clear that had the confrontation between the United States and the old USSR escalated beyond a limited one, the result would have been unthinkable: complete and utter obliteration in the Northern Hemisphere.

As we chatted online, I asked Sally something along the lines of: "Why should I go back and read *Warday* now?" I got a one sentence reply: "Check out pages 213–217." Now, admittedly, it was probably in the early 1990s when I had read *Warday*, so I had forgotten many of the specifics of the story. So, I checked out those pages. And what did I find? Well, I'll tell you what I found: a five-page chapter on a creature not at all unlike Mothman. This issue of a Mothman-type creature being associated with a devastating attack on the United States in fictional form (*Warday,* of course) eerily parallels what people were talking about in August 2017: dreams of a nuclear event and a tie-in with Mothman.

A couple of quotes from *Warday* will give you an idea of the nature of this aspect of the story. The title of the chapter is "Rumors: Mutants and Super-Beasts." We're told, under a heading of "Rumor": "There is a gigantic beast with bat wings and red, burning eyes that has attacked adults and carried off children. The creature stands seven feet tall and makes a soft whistling noise. It is often seen on roofs in populated areas, but only at night."

A further extract from Strieber's book concerns the testimony of an alleged eyewitness to the flying beast in California: "I had just gotten off the Glendale trolley when I heard this soft sort of cooing noise coming from the roof of a house. The sound was repeated and I turned to look toward the house. Standing on the roof was what looked like a man wrapped in a cloak. Then it spread its wings and whoosh! it was right on top of me."

"The center of the memory is that it was suddenly and completely destroyed by an atomic bomb, and nobody knows who detonated it."

It is important to note, in light of the Mothman-like references, that *Warday* is not a piece of wild science fiction. The story of the Mothman-type beast is only included in the book to demonstrate how, in the aftermath of the war, strange and bizarre rumors surface and spread among the survivors. I did, however, find it intriguing that *Warday* makes a connection between a nuclear war and "a gigantic beast with bat wings and red, burning eyes." This was, of course, what was being reported in mid-2017.

Interestingly, in 1995, Strieber himself had a graphic dream of a nuclear explosion that destroys Washington, D.C., in 2036. The event would see the end of the government as we know it today and the rise, in the wake of the disaster, of a full-blown dictatorship. In his 1997 book *The Secret School,* Strieber says of this dream—or perhaps brief view of what was to come via a future self—that "Washington, D.C. is in ruins. However, this isn't the center of the memory. The center of the memory is that it was suddenly and completely destroyed by an atomic bomb, and nobody knows who detonated it."

The final message on this topic came from Roger Pingleton, the husband of Jill S. Pingleton, who had contacted me a few days earlier. On August 28, Roger

TIME TRAVEL: THE SCIENCE AND SCIENCE FICTION

wrote: "Hi Nick. My wife informed me of the subject matter of some of your recent articles, and encouraged me to reach out with my experiences. Before Jill and I were married, I drove to Serpent Mound in Ohio on 11/11/2011 to meet with Jill and a group of people. I've always had weird feelings about the Ohio River valley, the mound building Indians, and the deities they worshiped there. And being that Serpent Mound is so close to Point Pleasant, I couldn't resist driving on to Point Pleasant, WV, after Serpent Mound. I slept a few hours in my pickup and drove on to Point Pleasant in the wee hours of the morning. I'd estimate my arrival to have been sometime between 3 a.m. and 4 a.m. The best I can describe the feeling I got driving through the back roads of WV is visceral. I felt like I was being called to be there at that early hour.

"Driving in the back roads near the old munitions facility, I saw, up ahead, two bright circular lights above the road. My thought was that they were circular taillights, that there was a hill up ahead and that I might catch up with the vehicle. The thing is, as I drove, I discovered there was no hill ahead of me, which freaked me out, because those two red lights were definitely higher than they should have been. Then I started thinking, 'What vehicle has round taillights these days?' A corvette maybe, but they have more than two.

"I couldn't help but think I had seen the eyes of the Mothman. I know it's weak evidence, but I can't come up with another plausible explanation. Not long after that, I had an apocalyptic dream. We live just south of Indianapolis. The city is on a grid and as such it's easy to tell directions. In my dream there was a giant explosion on the NW–SE diagonal axis. When I woke I worried that the city was Cincinnati, since it was to the SE of Indy, and my sister lives there, but then I realized it could also have been Chicago, which is to the NW of Indy. I didn't put these two events together until the Mothman sightings occurred. I truly hope I am wrong about these connections."

Of course, it's possible that all of these nightmares will not occur in our reality or even in our timeline. They may occur in one of many infinite other timelines.

I hoped that Roger Pingleton was wrong too. And I hope that just about everyone else who had such dreams was wrong. To be sure, it was a dark and tense period, with so many people having terrifying dreams of a worldwide, disastrous nuclear war with links to an assassination of a U.S. president. So far, those nightmares have not come true. I hoped—and still hope—they never will come true.

Of course, it's possible that all of these nightmares will not occur in our reality or even in our timeline. They may occur in one of many infinite other timelines—an issue I'll return to.

One final case that may have an impact on how dreams, nightmares, and ideas can cross over into future times is undeniably weird. On the matter of advance knowledge of the terrorist attacks of September 11, 2001, there is a curious and downright surreal story that has a link to none other than the television show *The X-Files*, which aired from 1993 to 2002 and then for two more seasons in 2016 and 2018. Although the characters Fox Mulder and Dana Scully were the primary focus of each episode as they strove to uncover the truth about a number of cosmic conspiracies, from time to time they received significant help from a trio of eccentric conspiracy theorists. The characters were John Byers, Melvin Frohike, and Richard Langly, who published the *Magic Bullet Newsletter*. The three characters became known as the Lone Gunmen. Such was the enthusiasm that the show's fans had for Langly, Frohike, and Byers that in early 2001 they were given their very own, short-lived series. The name of the series surprised no one: *The Lone Gunmen*.

The first episode aired on March 4, 2001. Its title was *Pilot*. The plot was chillingly similar to the events that went down later that year on September 11. In the show, a computer hacker takes control of a Boeing 727 passenger plane and flies it toward the World Trade Center with the specific intention of crashing the plane into one of the Twin Towers. It's only at the very last moment that the Lone Gunmen are able to hack the hacker and avert disaster and death for those aboard the plane and those inside the World Trade Center.

Was it prescient that in the pilot episode of the short-lived TV series *The Lone Gunmen*, a computer hacker takes control of a Boeing 727 passenger plane, intending to crash it into one of the Twin Towers?

The story gets even more intriguing: the hacker is not just some random, crazy guy. The plan is the work of a powerful group buried deep within the U.S. government. The secret plan, had it worked, was to put the blame for the World Trade Center attacks on one or more foreign dictators who are "begging to be smart-bombed" by the U.S. military. It should be stressed that there is no evidence that the creators of *The Lone Gunmen* had any advance knowledge of 9/11. It is worth noting, however, that there seemed to be a deep reluctance on the part of the media to address the storyline of *Pilot* and its parallels to 9/11—not to mention that the episode had its premiere broadcast in Australia just *thirteen days* before the events of September 11 occurred.

One of those who commented on this odd state of affairs was investigative journalist Christopher Bollyn. He said: "Rather than being discussed in the media as a prescient warning of the possibility of such an attack, the pilot episode of *The Lone Gunmen* series seemed to have been quietly forgotten. While an estimated 13.2 million Fox TV viewers are reported to have watched the pilot episode … , when life imitated art just six months later on 9/11, no one in the media seemed to recall the program."

Frank Spotnitz was one of the executive producers of *The Lone Gunmen*. He said: "I woke up on September 11 and saw it on TV and the first thing I thought of was *The Lone Gunmen*. But then in the weeks and months that followed, almost no one noticed the connection. What's disturbing about it to me is, you think as a fiction writer that if you can imagine this scenario, then the people in power in the government who are there to imagine disaster scenarios can imagine it, too."

Robert McLachlan was the director of photography on *The Lone Gunmen*. He, too, observed: "It was odd that nobody referenced it. In the ensuing press nobody mentioned that [9/11] echoed something that had been seen before."

Jeffrey King, who has carefully studied the 9/11 events, asked: "Is this just a case of life imitating art, or did [the production company] know something about the upcoming attacks? Was this an attempt to use the highly visible platform of the first episode of a new series (and a spin-off from the very popular *X-Files*) to make enough people aware of the scenario that it would become too risky to implement? Or was it just one of those ideas that was 'in the air' at the time, an expression of the zeitgeist?"

King had more to impart: "Great and traumatic events always seem to be preceded by certain foreshadowings, like the upstream standing waves that form behind a rock in the streambed. Perhaps this is just another in the endless string of odd synchronicities surrounding the events of 9/11, peculiar juxtapositions of events that must eventually strain the credulity of even the most devoted coincidence theorist, though no single one rises to the level of a smoking gun."

CHAPTER 10

STRANGERS IN THE SKIES

Since the 1980s, sightings of large, triangular-shaped UFOs, usually described as black in color, have been reported throughout the world. Observers frequently cite a low humming noise emanating from the crafts and mention rounded rather than angled corners. The sheer proliferation of such reports has led some ufological commentators to strongly suspect that the Flying Triangles (or FTs, as they have also come to be known) are prime examples of still-classified aircraft, the development of which was secretly begun in the 1980s by elements of the U.S. Department of Defense. Largely, UFO researchers are split into two camps. One says that the Flying Triangles are the creations of the American military, and the other claims they are flown by extraterrestrials. There is, however, a third theory for all of this mystery.

The late Omar Fowler, a UFO expert who died in 2017, had a theory that the Flying Triangles might be nothing less than the aerial vehicles of time travelers. Fowler had a good reason for that. On one occasion, in September 1992, over a large area of woodland in central England called the Cannock Chase, Omar secured a piece of testimony from a man named Alan Ball. He agreed to meet Ball at the Chase one morning. Incredibly, Ball claimed that he was taken aboard a huge, black Flying Triangle late one night while he was driving home. He was "beamed into" the craft and subjected to a series of medical experiments by three small, humanoid figures in what "looked like a medical lab."

Ten or 15 minutes into the experimentation, Ball claimed that his mind was flooded by images of the U.K. in ruins. It was a definitive Armageddon: buildings in all

Omar Fowler, a UFO expert who died in 2017, had a theory that so-called Flying Triangles might be the aerial vehicles of time travelers.

directions were destroyed, charred human bodies lay everywhere, and it was clear that a nuclear event had occurred. The assumption was that the event was not an isolated one—that the United Kingdom, and possibly everywhere else, had been destroyed. After having the terrifying vision, Ball was dropped off the craft, literally, from a door about five feet above the woods. Ball remained in a state of terror for several days.

Fowler was open to the idea that the Flying Triangles could be extraterrestrial in nature, but he wasn't able to shake off the disturbing possibility the craft were piloted by time travelers.

Particularly of interest is the fact that Fowler had in his files two extremely similar cases, also at the Cannock Chase woods, of people taken onto Flying Triangles and exposed to images in their minds of a massive apocalypse. Fowler was open to the idea that the Flying Triangles could be extraterrestrial in nature, but he wasn't able to shake off the disturbing possibility the craft were piloted by time travelers and that what Ball and the two other people were seeing was a future still to come—and that all three victims had been taken to the future and then returned to 1992. With that said, let's now take a deeper look at the phenomenon of the Flying Triangles.

In March 1993 a series of earth-shattering Flying Triangle encounters occurred in British airspace that went on to have a profound effect on high-ranking sources within both the Royal Air Force (RAF) and the Ministry of Defense (MoD). Importantly, it was that single wave of encounters that ultimately led senior military and defense personnel to liaise with their American counterparts to try to determine, once and for all, if the FTs were of terrestrial or extraterrestrial origin. The story comes from one of those at the forefront of the study into the aforementioned sightings: Nick Pope, who, for three years (1991–1994), officially investigated UFO incidents on behalf of the Ministry of Defense.

Long since retired from the MoD, Pope reveals his role in—and his knowledge of—the March 1993 UFO encounters over the United Kingdom. He says: "I arrived at the office at about 8.30 a.m. or 9.00 a.m. on the morning of March 31, 1993, and my telephone was ringing. I picked it up and there was a police officer on the other end making a UFO report. Now, he was based in Devon and told me an account of an incident that had taken place in the early hours of that particular day when he and a colleague who had been on night patrol saw a triangular-shaped UFO at fairly high altitude. He said that the motion was fairly steady and that there were lights at the edges with a fainter light in the middle.

"To me, this was already a description that was becoming quite familiar both from one or two reports that I'd received at the Ministry of Defense over the years and from my own study and research into the UFO literature. In other words, I was aware that this was a commonly reported shape for a UFO."

Pope continues: "I was also quite pleased to get a report from a police officer. I won't say that it was rare, but it was slightly unusual to have reports from trained observers like police and military. I would say that, of the reports I received in my time at the UFO desk, less than five per cent came from, collectively, pilots, military officers and the police. I had spoken, socially, to numerous Royal Air Force pilots who'd had personal sightings, but who had never reported them for fear of ridicule.

"But that police report was very much the first of many that came in that day and over the next week or so. When taken together, the sightings described took place in a range of times—the earliest was about 11–11.30 p.m. on the evening of the 30th and the latest was about 1.45 a.m. in the early hours of the 31st."

What was it precisely that made the police officer's report stand out? Pope explains: "He said to me: 'I've been on night patrols for years, but I've never seen anything like this in my entire life.' Well, reports such as this came through thick and fast over the course of the next week or so; more and more reports came in from police stations, the public and local RAF stations. In fact, I would say that the total number of reports easily exceeded one hundred."

It is clear from what Pope has to say that three reports in particular stood out from the others. The first concerned a family based in Rugeley, Staffordshire, England, who had viewed a remarkable aerial vehicle near the sprawling forest that is Cannock Chase. Pope reveals the facts:

TIME TRAVEL: THE SCIENCE AND SCIENCE FICTION

Various reports have been made of remarkable aerial vehicles near the sprawling forest of Cannock Chase, Staffordshire, England.

"This report was brought to my attention by the Community Relations Office at RAF Cosford [Shropshire]. The report had come direct from the family and sounded particularly interesting because, unlike some of the other sightings, this one was of an object flying at very low level. There had been a family gathering and several members of the family were out on the drive—really just saying goodbye to their relatives who were about to drive off. Suddenly, this large, triangular-shaped craft flew over them very, very slowly. This was a flat triangle, with a light in each corner and a larger light in the direct centre of the underside of the craft."

In fact, not unlike the report filed by the police reports from Devon?

"Exactly. But there was something else that I'd come across in my investigations that was also present in the Rugeley case," says Pope. "This was a low-frequency humming sound coming from the UFO; a humming that they actually described as being quite unpleasant. Imagine standing in front of the speakers at a pop concert and almost feeling the sound as well as hearing it—that was the effect that they reported. Well, they were so excited and overwhelmed that two of them leapt into the car to give chase!

"As they did so, they came to a point where they thought the UFO was so low that it must have come down in a nearby field. Well, they parked the car, jumped out and looked around. But there was absolutely nothing there; the UFO had gone."

The night's activities had barely begun, as Pope relates: "The two most significant reports began at RAF Cosford shortly after the encounter at Rugeley. This was definitely the highlight and was one of the best sighting reports I received in my entire posting. The report itself came from a guard patrol at Cosford. They were on duty manning entrance points, checking the perimeter fence and such like. All the members of the patrol saw the UFO and, again, the description was pretty much the same as most of the others. In this case, though, the UFO was at medium-to-high altitude."

Pope makes an important observation: "Remember that these witnesses were people who see in a normal course of business all sorts of aircraft activity, meteorites, fireballs and so on, and they considered it absolutely out of the ordinary.

"They didn't make a standard report: what they did was to submit an actual 2–3 page report which went up their chain of command and then the report was forwarded on to me. In that report, they stated that the UFO passed directly over the base and that this was of particular concern to them. They made immediate checks with various Air Traffic Control radar centres but nothing appeared on the screens. It was this factor that made them particularly keen to make an official report. This was at around 1.00 a.m."

Whatever the origin of the unknown vehicle, it appeared that its activities were far from over. Pope explains: "They noticed that this 'Flying Triangle' was heading on a direct line for RAF Shawbury, which is some twelve to fifteen miles on. Now, the main concern of the Cosford patrol was to alert Shawbury that the UFO was coming their way; but they also wanted confirmation that they weren't having a mass hallucination.

A Royal Air Force officer at the Shawbury base reported seeing a "massive, triangular-shaped craft" flying at a height of about 200 feet.

TIME TRAVEL: THE SCIENCE AND SCIENCE FICTION

"They took a decision to call Shawbury and this was answered by the Meteor-ological Officer. You have to realize that at that time there was literally just a skeleton staff operating, so the Meteorological Officer was, essentially, on his own. So, he took a decision to go outside, look in the direction of RAF Cosford and see what he could see.

"Sure enough, he could see this light coming towards him and it got closer and closer and lower and lower. Next thing, he was looking at this massive, triangular-shaped craft flying at what was a height of no more than two hundred feet, just to the side of the base and only about two hundred feet from the perimeter fence."

Bearing in mind the fact that the meteorological officer at RAF Shawbury could be considered a reliable witness and someone well trained in recognizing numerous types of aerial phenomena, was he able to gauge the size of the object? Pope says:

"Very much so: military officers are very good at gauging sizes of aircraft and they're very precise. His quote to me was that the UFO's size was midway between that of a C-130 Hercules and a Boeing 747 'Jumbo Jet.' Now, he had eight years' worth of experience with the Royal Air Force, and a Meteorological Officer is generally much better qualified than most for looking at things in the night sky. And there were other factors too: like the family in Rugeley, he heard this most unpleasant low-frequency hum; but unlike their experience, he saw the craft fire a beam of light down to the ground. He felt that it was something like a laser beam or a searchlight. The light was tracking very rapidly back and forth and sweeping one of the fields adjacent to the base.

"He also said—and he admitted this was speculation—that it was as if the UFO was looking for something. Now, the speed of the UFO was extremely slow—no more than twenty or thirty miles per hour, which in itself is quite extraordinary. As far as the description is concerned, he said that it was fairly featureless—a sort of flat, triangular-shaped craft, or possibly a bit more diamond-shaped. But if all the descriptions had been identical I would have been surprised."

Perhaps the most revealing aspect of the RAF Shawbury encounter was the way in which the object made its exit, as Nick Pope reveals. Still speaking of the meteorological officer, Pope explains: "He said that the beam of light retracted into the craft, which then seemed to gain a little bit of height. But then, in an absolute instant, the UFO moved from a speed of about twenty or thirty miles per hour to a speed of several hundreds of miles per hour—if not thousands! It just suddenly moved off to the horizon and then out of sight in no more than a second or so—and there was no sonic boom. Well, of course, when I received this report and the one from Cosford, I launched as full an investigation as I possibly could."

As Nick Pope makes abundantly clear, that investigation proved to be extraordinary, to say the least. "Even though it was fairly obvious to me that there were a number of things that this object was not, I still made the checks anyway to try to eliminate absolutely every possibility."

Pope also notes: "I had a feeling that this one was going to go right up the chain of command." He was not wrong.

He continues: "Checks were also made with various Air Traffic Control Centers, with Air Defense experts and Air Defense radar systems; and although at one point we thought we had caught the UFO on radar, it eventually turned out that there was nothing. After these checks were made and we were able to establish that the UFO hadn't been caught on radar, the Royal Air Force was quite interested. There isn't really a corporate view on UFOs; it really does go down to the belief of the individual. But, enough people realized that there was something exciting and out of the ordinary going on and they, too, got caught up in all that excitement."

Initially, suggestions were put forward that all of the sightings were simply the result of a satellite reentering Earth's atmosphere: "I spoke to the Space Information Officer at RAF Fylingdales; this is the Ballistic Missile Early Warning Centre. They've got very powerful space tracking radar that can pick up and track all sorts of objects at orbital heights. Now, they raised the possibility that we were looking at the re-entry of one of the Russian *Cosmos* satellites.

"Contrary to what some people have said, however, Fylingdales were very unsure as to whether or not the satellite would even have been visible from the U.K. at all during that time. But even if there was a re-entering satellite in the skies, it certainly couldn't explain the very close encounter at RAF Shawbury. Don't forget, too, that a satellite burn-up is very much like a meteor shower with a few tracks of light flashing across the sky. In this sighting, however, it was a case of one military base actually reporting to another and saying: 'It's coming your way....' So this rules out a satellite burn-out."

Sightings of Flying Triangles and other UFOs had occurred in the U.K. in Devon, Cornwall, South Wales, the Midlands, Southampton, and Yorkshire, as well as in Ireland, Belgium, and elsewhere in Europe.

Pope then took his investigation to another level.

"My next step was to get a map and plot out the various locations where the

UFO—or UFOs—had been seen. Well, that didn't work out. I was confronted with a map of haphazard sightings all around the country. There was certainly a concentration of sightings in Devon, Cornwall, South Wales and the Midlands. But there were also sightings from Southampton and Yorkshire; and I knew that there were reports from Ireland, Belgium and elsewhere in Europe. And these were just the tip of the iceberg.

"One interesting point that then occurred to me was that we were dealing with activity on exactly the same night—but three years later—to a very famous wave of sightings of very similar craft seen over Belgium. And my favorite theory about this or at least an idea I floated about—was that this was a deliberate move on the part of whoever was operating the craft."

"The earliest date that the story could have run would have been April 1—April Fools' Day! Again, a little indicator, perhaps, of an intelligence and possibly even some form of humor."

Pope explains his line of thinking: "For example, if the media had got a hold of this, it would have been too late to get it in the newspapers on March 31; so, the earliest date that the story could have run would have been April 1—April Fools' Day! Again, a little indicator, perhaps, of an intelligence and possibly even some form of humor."

Of course, it could be argued that this would serve as excellent cover if the Flying Triangle that was seen near RAF Shawbury was a terrestrial aircraft (albeit a distinctly secret one) as opposed to something extraterrestrial. Pope acknowledges this, saying: "We decided that we couldn't ignore the various rumors that were doing the rounds about a supposed Top Secret aircraft developed by the U.S. government and called *Aurora*—or, indeed, *any* hypersonic and/or prototype aircraft operated by the Americans.

"There had been persistent rumors in the aviation world and amongst the UFO lobby that the SR71 *Blackbird* had been replaced by a hypersonic aircraft code-named *Aurora* and that that was what the Flying Triangles really were. I was well aware that there had been some interesting stories about visual and radar sightings around certain air bases; however, I hadn't put much store in these rumors—not least because there had been some very definitive denials from the Americans.

"I know there's a lot of cynicism about government and the military. And although officialdom may refuse to answer a question and may sometimes give a misleading answer, outright lying is incredibly rare. And when it does occur, if it's uncovered it almost certainly leads to resignation.

"But with the March 1993 sightings—and in spite of the denials from the Americans that they were responsible for the Flying Triangles—we did contact them to make inquiries. This was because they have the responsibility pertaining to the U.S. presence in Britain. Those inquiries bore absolutely no fruit at all. The Americans said: 'No. We can shed no light at all on the UFO sightings that have led to your inquiry.'"

> *"As we were making our inquiries, they turned the question around and wanted to know if our Royal Air Force had a triangular-shaped, hypersonic prototype aircraft of some sort. So, presumably, the Americans were having Flying Triangle sightings, too."*

Pope is able to disclose, however, that the liaison with the Americans was not without its moment of intrigue. "If anything," he relates, "there was an interesting little hint that the Americans, too, were seeing these Flying Triangles over their territory. As we were making our inquiries, they turned the question around and wanted to know if our Royal Air Force had a triangular-shaped, hypersonic prototype aircraft of some sort. So, presumably, the Americans were having Flying Triangle sightings, too.

"But this was interesting, in light of the fact that the Americans supposedly got out of UFO investigations back in 1969 when the Air Force's Project Blue Book closed down. Of course, you may not officially be *in* the UFO game, but you are certainly going to be aware of—and take an interest in—reports of structured craft in your airspace. So, essentially, we drew blanks with the Americans."

At the time, Nick Pope and the secretariat of the air staff were not the only ones who were addressing the issue of whether the Americans were flying an *Aurora*-type craft in British airspace. In early 1995, for example, this issue was brought up in none other than the houses of Parliament. It was January 26, and the following exchange took place between Llew Smith MP and Nicholas Soames, at the time the minister for the Armed Forces:

Mr. Llew Smith: To ask the Secretary of State for Defense how many *Aurora* prototype aircraft of the United States Air Force are based at the Machrihanish Air Force Base in Argyll; and for what period permission has been given for basing these aircraft in the United Kingdom.

Mr. Soames: There are no United States Air Force prototype aircraft based at RAF Machrihanish and no authorization has been given by Her Majesty's Government to the United States Air Force, or any other U.S. body, to operate such aircraft within or from the United Kingdom.

TIME TRAVEL: THE SCIENCE AND SCIENCE FICTION

As this exchange made abundantly clear, even during a parliamentary debate nothing had surfaced to suggest—officially, at least—that the American government was in any way implicated in the mystery of the Flying Triangles. Back to Nick Pope:

"Bearing in mind that the Americans had inquired—at an official level, no less—if the British Royal Air Force had in its employ something broadly fitting the description of a Flying Triangle, and we had said 'No,' I still felt obliged to address the issue of whether or not the rumors about secret aircraft being flown by us were true.

"First, from my own knowledge of prototype aircraft, un-manned aerial vehicles and so on, the Triangles don't fit into the typical pattern, and I'll explain why. Where we do have such pieces of kit, they're not tested over the heads of 'Joe Public'; they're tested in a small number of clearly defined ranges and danger areas—mostly out at sea such as the Abberporth Range in Cardigan Bay.

"You simply do not fly a prototype craft over a military base or over the centre of Rugeley or wherever."

"You simply do not fly a prototype craft over a military base or over the centre of Rugeley or wherever, and run the risk that someone will either (a) scramble a [Tornado] F-3 [aircraft] to try and intercept it; or (b) take a photograph of it which will end up on the front page of *The Sun* or *Jane's Defense Weekly*. It's simply not the way that things are done.

"We checked domestically anything that might have been flying. But if we'd have been poking our noses into something that didn't concern us, the investigation would have been quietly switched off. In fact, the opposite happened. We were making big waves throughout the Royal Air Force, the Ministry of Defense and at an international level. So the domestic secret aircraft theory is interesting but it doesn't hold water."

Three years prior to the extraordinary events at RAF Cosford and RAF Shawbury, similar objects were seen on repeated occasions in Belgian airspace in 1989 and 1990. In view of this, was any form of approach made to the Belgian military to ascertain their views?

"Yes," says Pope. "I approached the Belgians to get a comparison after their sightings. I phoned the Air Attaché at the British Embassy in Brussels and he spoke to one of the F-16 pilots who had been scrambled to intercept a Flying Triangle over Belgium back in 1990. Well, the Air Attaché reported back to me that the corporate view of the Belgian Defense Staff was that they did believe that they were dealing with a solid, structured craft.

"Apparently, the word from the Belgians was: 'Thank God it was friendly.' If it hadn't been, it was made clear to me that there was very little that the Belgian Air Force could have done anyway—despite the fact that the F-16 is no slouch."

TIME TRAVEL: THE SCIENCE AND SCIENCE FICTION

With the secret weapon angle disposed of as far as Nick Pope was concerned, what was his next step in the investigation?

"There was only one place to go and that was up the chain of command, and I briefed my head of division. He was notoriously skeptical about UFOs and generally made no secret of the fact that he thought that it was all a waste of time and resources. But he had been quite impressed by the Shawbury and Cosford events—even to the point of making some attempts to plot the course of the UFO.

"In fact, I recall him bounding into the office in a state of some considerable excitement when he thought that he had found indications of a straight-line track. I had copied some of the reports; but what he didn't have was a batch of reports that had just come in and that painted a totally different picture.

"Well, I just thought that this needed to go up the chain of command. The main addressee was the Assistant Chief of the Air Staff; so what I did was to summarize the events on a couple of sides of paper and attach the original reports—the typed report submitted by the patrol sighting at Cosford and my own write-up of the RAF Shawbury sighting.

"He took a few days to have a look at all the paperwork and then passed it back down the chain of command with a message that said: 'This is extremely interesting. It is a genuine mystery but clearly you've made all the checks that we could reasonably make and it's difficult to see how we can take this any further.' And that was essentially where the matter rested."

Today, does Nick Pope feel that the assessment of the assistant chief of the air staff was a fair one?

"Well, yes and no. I felt extremely uncomfortable that we had a clear breach of the U.K. air defense region; and we had two Royal Air Force bases pretty much being over-flown by a structured craft and yet we had nothing on radar and absolutely no explanation. I applied our own standard line on UFOs and asked myself the questions: Is this of no defense significance? What if the craft had been hostile? What if a bomb-bay had suddenly opened up and it had attacked these bases? If that had been the case, and with the UFO not appearing on radar, the first we would have known would have been when the bombs were falling. So, I came to the conclusion that this was of extreme defense significance.

"I'm naturally suspicious of anyone that doesn't declare their hand. And although there may be some very good reasons for them remaining covert, I think that from a military and defense point of view, you have to say that there is a potential UFO threat."

TIME TRAVEL: THE SCIENCE AND SCIENCE FICTION

CHAPTER 11

HEADING INTO THE PAST

P auline Charlesworth had a very strange experience back in the 1980s, which just might have involved Pauline being torn out of our time and into the past—and not the recent past but several thousand years ago. The location was an ancient hillfort, as it is called, in Cannock, Staffordshire, England, known as Castle Ring. Before I get to the story, it's important that I share with you the history of this undeniably magical locale. The website *Historic England* says: "The monument is situated at the south eastern edge of Cannock Chase and includes the earthwork and buried remains of an Iron Age hillfort and the ruins and buried remains of a small medieval building identified as part of a hunting lodge. Castle Ring occupies the summit of a small hill which forms the highest point on the Chase. The hillfort is an irregular pentagon in plan and its multiple defenses enclose an area of c. 3.6ha. For the majority of their circuit these defenses include a sequence of banks and ditches, usually three banks with two ditches in between them. On the east side of the hillfort, where the approach to the site is over more level terrain, the central bank is more substantial, the outer line of defenses projects outwards and an additional length of ditch and counterscarp banking has been added."

Now let us surf our way into the past and to a time when Castle Ring was very different.

According to Pauline, it was a bright, summery day in July 1986 when her strange encounter occurred. As she worked on Saturdays, Pauline explained to me when we first met, she had a regular day off work during the week, and she had chosen this particular day to prepare a picnic basket and take a trip up to Castle Ring. On arriving, she prepared for herself a comfortable place to sit, stretched out a blanket on the ground, and opened up her picnic basket containing drinks, fruit, and sandwiches. For more than an hour she sat and read a book, but then something very curious happened. It was as if, Pauline explained, she was sitting within the confines of a vacuum and all of the surrounding noises, such as the birds whistling and the branches of the trees gently swaying, stopped completely. Pauline also said that "what was there wasn't quite right." She explained: "The best way I can describe it is to say it was like I wasn't really on the Chase, but it was as if I was in someone's dream of what the Chase should look like; as if it was all a mirage, but a good one."

> *It was as if, Pauline explained, she was sitting within the confines of a vacuum and all of the surrounding noises, such as the birds whistling and the branches of the trees gently swaying, stopped completely.*

Then, out of the trees came a horrific form running directly toward her. It was, said Pauline, a man, but one quite unlike any she had ever seen before. He had long, filthy hair, a matted beard, and a "dumpy" face that she described as more prehistoric than modern in appearance. He was relatively short in height, perhaps no more than five feet two inches, and was clad in animal skins that extended from his waist to his knees, with a long piece of animal skin draped over his right shoulder. In his right hand, the man held what were undoubtedly the large antlers of a deer that had been expertly fashioned into a dagger-like weapon that looked like it could inflict some serious damage.

Pauline said that it was very difficult to ascertain who was more scared, her or the man. While she stared at him in stark terror, he eyed her curiously and in what Pauline described as a disturbing and sinister fashion. On several occasions he uttered what sounded like the words of an unknown language. "It was like he was angry and firing questions at me," she said. But that was not all.

In the distance, Pauline could hear other voices getting ever closer, a sound that grew into an impressive crescendo. Soon she identified the source of the noise: through a break in the trees came perhaps 30 or 40 similarly clad people, mostly men but a few women, all chanting in an unknown, and presumably ancient, tongue.

It was soon made clear to Pauline that some sort of significant ceremony was about to take place inside Castle Ring, and that she was right in the heart of the brewing action. The men and women proceeded to sit down at the edges of the ring. One man, much taller than the rest and presumably the "leader of the group," as Pauline said, marched over to her and said something wholly unintelligible, although she understood by the curt wave of his arm that he meant for her to get out of the circle. This she did quickly and retreated with shaking legs to the tree line. For more than 15 minutes, she sat transfixed with overwhelming terror as this curious band of people continued to chant and sway in rhythmic, hypnotic fashion.

TIME TRAVEL: THE SCIENCE AND SCIENCE FICTION

At Castle Ring, the Iron Age ruins of a hill fortress in Cannock, Staffordshire, England, Pauline Charlesworth may have been torn out of the present and launched into the distant past.

Suddenly, out of the sky came the most horrific thing that Pauline had seen in her entire life. It was, she recalled, a creature about four feet in height, human in shape, with oily, greasy black skin, thin arms and legs, and a pair of large, bat-like, leathery wings. Just for good measure, it had two hideous, glowing red eyes, too. "It was like the devil," recalled Pauline, with understandable justification.

The creature slowly dropped to the ground and prowled around the ring for a minute, staring at one and all and emitting hideous, ear-splitting shrieks. Suddenly, seven or eight of the men pounced on the creature, wrestled it to the ground, and bound it firmly with powerful ropes. It writhed and fought to get loose and tore into the flesh of the men with its claws, but it was finally subdued and dragged into the forest by the tribe members. The remainder of the party followed. After this, Pauline said, the strange atmosphere began to lift, and the area eventually returned to its original normality. For several minutes she stood her ground, too afraid to move, but finally returned on still-unsteady legs to her blanket and quickly scooped up both it and her picnic basket and ran to her car.

I have left until last one factor that, in some fashion, simply has to be connected to this saga. The book that Pauline had taken with her to read at Castle Ring was Robert Holdstock's acclaimed fantasy novel *Mythago Wood*. Interestingly, the book is one of my all-time favorites, and I read it at least once a year. The story tells how, after the end of World War II, one Steve Huxley returns to England upon receiving the news that his father has died. George Huxley had devoted his life to the exploration of the ancient Ryhope Wood that backed up against the family home and kept detailed records of his research into the mysterious area. But Ryhope Wood is unlike any other. It is inhabited by the "mythagos" of the book's title. And what, exactly, might they be? They are creatures and characters from British folklore and mythology, such as Robin Hood and King Arthur, whose curious existence is directly tied to the imaginations and minds of those who believe in them and who, in unconscious, collective fashion, help bring them to some form of quasi-independent life in the depths of the magical Ryhope Wood.

Of course, in view of the *Mythago Wood* connection, a skeptic might say that Pauline's unearthly experience was merely the result of a bizarre dream—or perhaps the worst of all nightmares. And maybe that really *is* all it was. More than a quarter of a century later, however, Pauline herself is still convinced that something very strange and diabolically evil occurred on that summer day in long-gone July 1986, and that she was provided with a unique glimpse into Staffordshire's very ancient past via time travel. Notably, there is a fascinating account that, to some degree, ties in with the experience of Pauline Charlesworth. It, too, fires someone wildly into the world of the past.

A respected authority on prehistory, R. C. C. Clay had just such an experience while driving at Bottlebrush Down, Dorset, England, an area strewn with old earthworks, during the winter of 1924. The story, however, did not surface until 1956. That was when Clay decided to share the details with an authority on all things ghostly and spectral, James Wentworth Day, an unlikeable, racist homophobe with fascist leanings, who penned such titles as *Here Are Ghosts and Witches*, *A Ghost Hunter's Game Book*, *In Search of Ghosts*, and *They Walk the Wild Places*.

The location of the extraordinary event that Clay related to an enthralled Wentworth Day was the A3081 road, between the Dorset villages of Cranborne and Sixpenny Handley, on farmland known locally as Bottlebrush Down. It was while Clay was driving home after spending a day excavating in the area, as the daylight was giving way to the magical, twilight hours, that he encountered something extraordinary. Maybe even *beyond* extraordinary.

At a point where the new road crossed with an old Roman road, a horseman, riding wildly and at high speed on the back of a huge and muscular stallion, seemingly appeared out of nowhere. But there was something wrong about this man, something terribly wrong. In Clay's very own words to a captivated Wentworth Day: "I could see that he was no ordinary horseman, for he had bare legs, and wore a long loose cloak. His horse had a long mane and tail, but I could see neither bridle nor stirrup. His face was turned towards me, but I could not see his features. He seemed

R. C. C. Clay, a respected authority on prehistory, encountered a horseman from the Bronze Age at a crossroads at Bottlebrush Down, Dorset, England.

to be threatening me with some implement, which he waved in his right hand above his head."

It is deeply fortunate that the witness in this case was Clay, a man with an expert and profound knowledge of English history, folklore, and times and people long gone. There was no doubt in Clay's mind that, having kept the rider in careful sight for around 300 feet, he could firmly identify the rider by his clothing and weapon as nothing less than a denizen of the Bronze Age. Incredibly, this would have placed his origins at some point between 2100 and 750 B.C.E. Not surprisingly, with darkness falling fast, Clay floored the accelerator and headed for home, somewhat shakily but decidedly excited, too.

His interest highly piqued, Clay began to make careful and somewhat wary inquiries in the area—feeling understandably somewhat tentative about the subject— to determine whether anyone else had ever seen the ancient hunter of the Downs. As it transpired, some people actually had. An old shepherd who had worked in the fields his whole life answered Clay's query by saying: "Do you mean the man on the horse who comes out of the opening in the pinewood?" When an amazed and excited Clay replied "Yes!" and asked further questions, it became clear to him that he was not the only person to have seen the enigmatic old rider of the land. A couple of years later, while still investigating the strange affair, he learned of yet another encounter with the ghostly man and horse. In this case, the witnesses were two girls, cycling from Sixpenny Handley to a Friday night dance at Cranborne, who were plunged into a state of fear by the presence of what sounded like the very same character encountered by Clay back in 1924.

As Clay told Wentworth Day in 1956, he knew of no more recent encounters with the horseman, but he theorized that what he had been fortunate enough to see was the spirit form of a Bronze Age hunter and his horse, both of whom had probably died under violent circumstances on the Downs and who—for a while, at least—roamed the very same hunting grounds that they had called home during their clearly turbulent physical lives.

There is an intriguing afterword to all of this: Clay was of the opinion that the man and his horse were specters—ghostly figures from times long gone. Perhaps they were. It should not be forgotten, though, that Clay said the horseman "seemed to be threatening me with some implement, which he waved in his right hand above his head."

Clay's statement was a very important, albeit brief, one. Here's why: *The horseman could clearly see Clay.* That much is clear from the fact that the ancient warrior was all tooled up and ready to go head to head with Clay, using what may have been a club or a blade. In many cases of ghostly activity, the spirits of the dead seem to repeat, over and over again, the very same actions, and with no sense of self-awareness. It's a situation akin to a never-ending, old-style tape loop being played. Clay's experience, however, was nothing like that. There are also these notable words from Clay on the matter of the fierce rider: "His face was turned towards me." This is further evidence that the two had each other in their sights. This makes me strongly suspect that Clay, the horseman, and the faithful steed had all briefly crossed time zones in an incredible fashion. Which of them was most baffled by the whole affair is anyone's guess.

Could it be that both Pauline Charlesworth and R. C. C. Clay stumbled into ages long gone via time travel? It's something we should not dismiss. Of course, the important question is: How did they manage to surf the centuries—and quite out of the blue? The answer might be: On occasion, we can travel completely randomly and in ways that we have no understanding of. Yet.

Having extensively investigated such reports, cryptozoologist and paranormal investigator Neil Arnold told the story of a friend named Corriene who had a mysterious encounter in 2003 at Kit's Coty House, a set of ancient standing stones in

Kent near a village called Blue Bell Hill. On one occasion near the stones, a sort of man-beast appeared a few hundred yards from her, "striding rather aggressively." It was more than six feet tall and "appeared to have a loin cloth around its waist and furred boots." Neil wondered if this might be a sort of time-traveling Neolithic hunter, an idea that intrigued Corriene, who had seen other such people, armed with spears, in the same vicinity on another occasion.

> *"She turned slowly and to her amazement saw a massive figure, standing over seven feet tall, seemingly covered in hair, just a few yards away."*

Significantly, this case did not stand alone, as Neil told me: "In 1974 a woman named Maureen—who I became very close friends with after working with her 10 years ago—was with her boyfriend in woods off a close named Sherwood Avenue, in Walderslade, which, as the crow flies, is only a couple of miles short of Blue Bell Hill. Her boyfriend was tending to the fire when Maureen felt as if she was being watched. She turned slowly and to her amazement saw a massive figure, standing over seven feet tall, seemingly covered in hair, just a few yards away. The figure had bright eyes which blinked. The figure seemed to crouch down behind the bushes very slowly, and Maureen asked her boyfriend if they could leave the area, and they did. Maureen never told her boyfriend what she'd seen. They eventually married and

Lustleigh Cleave, a valley near the charming village of Lustleigh in England's Dartmoor National Park, appears to be a "window area" or portal to paranormal events and other times.

TIME TRAVEL: THE SCIENCE AND SCIENCE FICTION

were together for many years. In fact, she never told anyone about what she'd seen until she realized what I was interested in."

Possibly of relevance to the theory and observations of Neil Arnold—and Maureen, too—is the story of the noted Devon folklorist and acclaimed writer Theo Brown. The author of such titles as *Devon Ghosts* and *Family Holidays around Dartmoor*, Brown collected a number of stories highly similar to those cited by Neil, including one chilling recollection by a friend of hers who had been walking alone at dusk near the Neolithic earthworks at the top of Lustleigh Cleave, which sits on the extreme east side of Dartmoor, in the Wrey Valley.

Lustleigh Cleave is an extraordinarily strange place at the best of times, and it appears to be a "window area" or portal as described by researcher and author Ronan Coghlan—who we'll come to in a later chapter on Bigfoot—where an inordinate number of unexplained incidents and anomalous phenomena seem to take place on an amazingly regular basis. Moreover, the remains of prehistoric stone huts can be seen in the direct vicinity, and an ancient burial monument, Datuidoc's Stone—which is estimated to have originated at some point around 550 to 600 C.E.—stands to this very day, pretty much as it did all those thousands of years ago.

Jonathan Downes, of the United Kingdom's Centre for Fortean Zoology, has this to say about the potentially time-traveling apparitions: "I have got reports of sightings of a ghostly Tudor hunting party, of mysterious lights in the sky, and even the apparitions of a pair of Roman Centurions at Lustleigh Cleave." Getting to the most important aspect of the story, the creatures apparently from another time, Downes says: "Theo Brown's friend saw, clearly, a family of 'cave men,' either naked and covered in hair or wrapped in the shaggy pelts of some wild animal, shambling around the stone circle at the top of the cleave."

CHAPTER 12

USING THE MIND TO HEAD INTO THE PAST

T he CIA takes a trip to Mars? And to the distant past? You may think it's a joke or the theme for a big-bucks sci-fi movie, but it's not. The CIA did exactly that in May 1984. They did so using the power of the human mind—specifically what is known as remote viewing. We'll begin in this chapter with the history of remote viewing. Then we'll get into the matter of Mars in the distant past. For decades, numerous nations worldwide have done their utmost to try to harness the mysterious powers of the mind and utilize them as tools of espionage. Extrasensory perception (ESP), clairvoyance, precognition, and astral projection have all been utilized by the CIA, the KGB, and British intelligence on more than a few occasions. As astonishing as it may sound, the world of "psychic 007s" is all too real. It's a subject that has been researched, with varying degrees of success, for decades. Not only a tool for spying in the present, the phenomenon of remote viewing also allows for travel into different realms of time.

The earliest indication of serious interest on the part of the U.S. government in the field of psychic phenomena can be found in a formerly classified CIA document written in 1977 by Dr. Kenneth A. Kress, an engineer at the time with the CIA's Office of Technical Services, titled "Parapsychology in Intelligence."

Not only a tool for spying in the present, the phenomenon of remote viewing also allows for travel into different realms of time.

TIME TRAVEL: THE SCIENCE AND SCIENCE FICTION

According to Kress: "Anecdotal reports of extrasensory perception capabilities have reached U.S. national security agencies at least since World War II, when Hitler was said to rely on astrologers and seers. Suggestions for military applications of ESP continued to be received after World War II. In 1952, the Department of Defense was lectured on the possible usefulness of extrasensory perception in psychological warfare.

"In 1961, the CIA's Office of Technical Services became interested in the claims of ESP. Technical project officers soon contacted Stephen I. Abrams, the Director of the Parapsychological Laboratory, Oxford University, England. Under the auspices of Project ULTRA, Abrams prepared a review article which claimed ESP was demonstrated but not understood or controllable."

Kress added: "The report was read with interest but produced no further action for another decade."

Finally, in the early 1970s the research began in earnest. In April 1972, Dr. Russell Targ, a laser physicist with a personal interest in parapsychology and the power of the human mind, met with CIA personnel from the Office of Strategic Intelligence specifically to discuss paranormal phenomena.

Of paramount concern to the CIA was Targ's information that the Soviet Union was deeply involved in researching psychic phenomena, mental telepathy, and ESP. It did not take the CIA long to realize that the purpose of the Soviet research was to determine whether ESP could be used as a tool of espionage. As one CIA agent said: "Can you imagine if a bunch of psychic 007's from Russia could focus their minds to short-circuit our missile systems or our satellite surveillance equipment and get access to classified information in this way? The possibilities—if it worked— would be disastrous."

It was this realization that galvanized the CIA into action. As the Kress report stated, in 1973 "the Office of Technical Services funded a $50,000 expanded effort in parapsychology."

The initial studies utilized a variety of people who were carefully and secretly brought into the project and who demonstrated a whole range of seemingly paranormal skills. Those same skills could not be reliably replicated on every occasion, however.

As evidence of this, Kenneth Kress informed his superiors: "One subject, by mental effort, apparently caused an increase in temperature; the action could not be duplicated by the second subject. The second subject was able to reproduce, with impres-

In 1973 the CIA's Office of Technical Services devoted $50,000 to studying parapsychological techniques such as remote viewing to "travel" to faraway places and times.

sive accuracy, information inside sealed envelopes. Under identical conditions, the first subject could reproduce nothing."

Similarly, some government-sponsored psychics in the period from 1973 to 1974 located secret missile installations in the Soviet Union, found terrorist groups in the Middle East, and successfully remote viewed the interior of the Chinese Embassy in Washington, D.C. Others, meanwhile, provided data that was sketchy and, at times, simply wrong.

It was the continuing rate of success versus the frequency of failure that led to heated debate within the CIA about the overall relevance and validity of the project.

In *Parapsychology in Intelligence,* Kenneth Kress confirmed this. After the CIA's remote-viewing team attempted to broaden the range of its operation and secure extra funding in mid-1973, said Kress, "I was told not to increase the scope of the project and not to anticipate any follow-on in this area. The project was too sensitive and potentially embarrassing."

Despite this, the CIA's research continued, with many of its advances due to a skilled psychic named Pat Price, who had achieved a number of extraordinary successes in the field of ESP, including successfully remotely viewing a sensitive installation that fell under the auspices of the National Security Agency and psychically penetrating missile sites in Libya.

Price's sudden and untimely death from a heart attack in 1975 indirectly led the CIA—according to the official story, at least—to minimize its research into psychic espionage.

Tim Rifat, who has deeply studied the world of top-secret governmental research into psychic spying, said of Pat Price's death: "It was alleged at the time that the Soviets poisoned Price. It would have been a top priority for the KGB to eliminate Price as his phenomenal remote-viewing abilities would have posed a significant danger to the USSR's paranormal warfare buildup. He may also have been the victim of an elite group of Russian psi-warriors trained to remotely kill enemies of the Soviet Union."

The scenario of research being minimized in the aftermath of Price's potentially suspicious passing was reinforced when, in 1995, a CIA-sponsored report titled "An Evaluation of the Remote-Viewing Program: Research and Operational Applications" was produced by the American Institutes for Research (AIR). In essence, the report stated that from an espionage and intelligence-gathering perspective, remote viewing and related phenomena were largely useless.

Not everyone agreed with that conclusion, however, including W. Adam Mandelbaum, author of *The Psychic Battlefield* and a former U.S. intelligence officer, who said: "The AIR report was US-intelligence-purchased disinformation intentionally formatted to misrepresent the true states of remote-viewing research, and the true operational utility of the phenomenon."

Regardless of whether the CIA's role in remote-viewing operations was downsized, terminated, or simply hidden from prying eyes, it is a matter of fact that additional agencies within the U.S. government, military, and intelligence community took—and continue to take—a deep interest in psychic espionage.

The Defense Intelligence Agency (DIA), for example, has had long-standing involvement and interest in understanding and using paranormal powers both on the battlefield and in the cloak-and-dagger world of espionage.

"A spy would be hypnotized, then his invisible 'spirit' would be ordered to leave his body, travel across barriers of space and time to a foreign government's security facility, and there read top-secret documents and relay back their information."

A 1972 report from the Defense Intelligence Agency, which specializes in military intelligence, claimed that by the late the 1970s Soviets would be able to "use ESP to steal the secrets of their enemies."

As an illustration of this, a DIA report from 1972 titled "Controlled Offensive Behavior: USSR" made an astonishing claim: "Before the end of the 1970s, Soviet diplomats will be able to sit in their foreign embassies and use ESP to steal the secrets of their enemies. A spy would be hypnotized, then his invisible 'spirit' would be ordered to leave his body, travel across barriers of space and time to a foreign government's security facility, and there read top-secret documents and relay back their information.

"The Soviets," the report continued, "are at least 25 years ahead of the U.S. in psychic research and have realized the immense military advantage of the psychic ability known as astral projection (out of the body travel)."

Similarly, in 1973 and 1975, the DIA commissioned two lengthy reports that delved deep into the heart of Soviet research of psychic phenomena and included details of one extraordinary experiment undertaken by the Russian military in the 1950s.

A somewhat disturbing extract from the DIA's files on this particular experiment states: "Dr. Pavel Naumov conducted animal bio-communication studies between a submerged Soviet Navy submarine and a shore research station. These tests involved a mother rabbit and her newborn litter and occurred around 1956."

The author of the report continued: "According to Naumov, Soviet scientists placed the baby rabbits aboard the submarine. They kept the mother rabbit in a laboratory on shore where they implanted electrodes in her brain. When the submarine was submerged, assistants killed the rabbits one by one. At each precise moment of death, the mother rabbit's brain produced detectable and record-able reactions."

Soviet experiments into ESP included putting a mother rabbit in an onshore laboratory and her babies in a submarine, then killing the babies and noting changes in the mother rabbit's brain.

TIME TRAVEL: THE SCIENCE AND SCIENCE FICTION

The DIA also noted: "As late as 1970 the precise protocol and results of this test described were believed to be classified."

Nevertheless, the DIA was able to determine that the Soviets' reasoning behind such experimentation was to try to understand the nature of ESP, astral projection, and the power of the mind—and even the existence of a soul—in animals such as dogs, rabbits, and primates. And if eventually understood in the animal kingdom, said the DIA, the Soviets' next step would be to focus on human beings and how those same phenomena might be used as a weapon of war and espionage.

In Britain the situation was broadly similar. Formerly classified files at the National Archives in Kew, Greater London, dating to the height of World War II reveal that elements of the British police occasionally and stealthily employed the use of dowsers—normally associated with searches for underground water—to locate victims buried under the rubble of inner-city destruction wrought by Nazi bomber pilots.

Such was the controversy surrounding this unique brand of psychic police work that even the government's wartime Ministry of Home Security became embroiled in the affair, urging caution in expressing "support for the mysterious" at such a "particularly dangerous time"—this despite the apparent success of its "dowsing detectives."

Still on the matter of Britain's secret spies, there is the matter of a purported novel titled *The Psychic Spy*. Written by Irene Allen-Block in 2013, it contains the following endorsement from me:

> In late 1970s London, a young woman is secretly recruited to work for British Intelligence. Her world soon becomes dominated by psychic-spying, enemy agents, assassinations, and suspicious deaths. Add to the mix the Lockerbie tragedy, the Falklands War, and the classified world of MI6, and you have a great story filled with adventure, intrigue and shadowy characters. As Irene Allen-Block skillfully shows, the mind is a mysterious and dangerous tool.

The publisher of the book, Glannant Ty, notes: "*The Psychic Spy* tells the story of Eileen Evans, a beautiful young woman and talented psychic who is unwittingly recruited by MI6 to join their new top secret Remote Viewing program 'Blue Star' during the heart of the Cold War in the 1970s and 80s. Eileen quickly finds herself embroiled in excitement and danger as she becomes a 'psychic spy' for British Intelligence. Finding forbidden love with another agent, Eileen descends into a dark world filled with political intrigue, danger and death. Not only must she cope with the possibility of losing her life, she must also struggle with the very real threat of losing her soul.

TIME TRAVEL: THE SCIENCE AND SCIENCE FICTION

Based on a true story, *The Psychic Spy* relates the adventures of a beautiful young woman and talented psychic who is unwittingly recruited by MI6 to join its top-secret remote-viewing program.

"Smart, sexy and filled with humor and peril, *The Psychic Spy* is a thrilling adventure that explores a little-known but very real world where governments use actual psychics to spy on their enemies, and in some cases, even their allies! Using her own real-life experiences as a remote viewer, Irene Allen-Block has created a powerful tale that should entertain and educate readers on a piece of history that has been hidden in the shadows."

In November 2001 it was revealed that the FBI had quietly approached private remote-viewing companies with a view to predicting likely targets of future terrorist attacks.

The Psychic Spy is made all the more intriguing by the fact that the book is actually a thinly veiled version of the real-life exploits of the author while, from the late 1970s onward, she was in the secret employ of British intelligence in the field of psychic spying.

What of today's world: Are psychic spies engaged in helping to end the War on Terror? In November 2001 it was revealed in the media that the FBI had quietly approached private remote-viewing companies with a view to predicting likely targets of future terrorist attacks.

As Lyn Buchanan, the author of *The Seventh Sense*—a book that examines Buchanan's personal role in the U.S. government's remote-viewing story—says on this subject: "We want the message to get to terrorists everywhere that no one attacks our country and kills our people and gets away with it. We can, and we will, find you."

Decades after official research began into remote viewing and ESP, it seems that the worlds of the psychic and the spy continue to cross paths.

CHAPTER 13

TIME TRAVELING TO
MARS, THE RED PLANET

I n 1984, behind closed doors, agents of the Central Intelligence Agency embarked on an ambitious project to try to "remote view" the planet Mars, specifically to find evidence of life on the planet. What is more, they were not trying to view Mars as it is today but as it was in the distant past. In other words, agents of the CIA were about to time travel into the past, using the power of the human mind to do so. Incredibly, they achieved their goal. The CIA took a trip into the past.

In other words, agents of the CIA were about to time travel into the past, using the power of the human mind to do so.

For those who may not know, remote viewing is a process by which a psychic individual focuses his or her mind on a particular location and describes its appearance and whether it exists now or existed in the past—years, centuries, or even millennia ago. It was on May 22, 1984, that the ambitious program began. While some of the pages declassified under the terms of the Freedom of Information Act are heavily redacted for national security reasons, others give us at least some sense of what was afoot behind the closed doors of the CIA headquarters in Langley, Virginia.

The time frame that the CIA had in mind for viewing Mars was approximately one million years B.C.E. Why? The CIA isn't telling. The remote viewer—whose name is deleted from the files—stated, "I kind of got an oblique view of a, ah, pyramid or pyramid form. It's very high, it's kind of sitting in a large depressed area. I'm tracking severe, severe clouds, more like dust storm[s]. I'm looking at an after-effect of a major geologic problem. I just keep seeing very large people. They appear thin and tall, but they're very large, wearing some kind of clothes."

The remote viewer then focuses on certain massive structures on Mars that appeared to have been built by intelligent beings: "Deep inside of a cavern, not of a cavern, more like [a] canyon. I'm looking up, the sides of a steep wall that seem to go on forever. And there's like a structure … it's like the wall of the canyon has been carved. Again, I'm getting a very large structures [sic] … huge sections of smooth stone … it's like a rabbit warren, corners of rooms, they're really huge. Perception is that the ceiling is very high, walls very wide.

"They have a … ah … appears to be the very end of a very large road and there's a marker thing that's very large. Keep getting Washington Monument overlay, it's like an obelisk … see pyramids … they're huge. It's filtered from storms or something…. They're like shelters from storms.

"Different chambers … but they're almost stripped of any kind of furnishings or anything, it's like ah … strictly [a] functional place for sleeping or that's not a good word, hibernations, some form, I can't, I get real raw inputs, storms, savage storm, and sleeping through storms.

***Does this specific piece of exceptional CIA
documentation, acquired via a psychic form of time
travel, tell of an ancient Martian civilization?***

In 1984 the CIA began an ambitious project to "remote view" the planet Mars, not as it exists today but as it was in the distant past, to find evidence of life on the planet.

TIME TRAVEL: THE SCIENCE AND SCIENCE FICTION

"They're ancient people. They're dying. It's past their time or age. They're very philosophic about it. They're just looking for a way to survive and they just can't. They're … ah … evidently was a … a group or party of them that went to find … ah … new place to live. It's like I'm getting all kinds of overwhelming input of the corruption of their environment. It's failing very rapidly and this group went somewhere, like a long way to find another place to live."

Does this specific piece of exceptional CIA documentation, acquired via a psychic form of time travel, tell of an ancient Martian civilization, one whose world spiraled into ecological collapse and forced them to a new world to live on? Maybe our Earth? It just might. The idea that Mars may once have been home to a race of advanced Martians, who fled their dying world when even they were unable to prevent an all-encompassing disaster, may well tie in with the strange story of what has become known as the Face on Mars, as we shall now see.

One person who was fascinated by all this was the late Mac Tonnies, whose book on these topics—*After the Martian Apocalypse*—makes for required reading. Most of Tonnies's work revolved around one particular area on Mars, a region called Cydonia. It is right in the heart of Cydonia that there can be found what appears to be a massive, human-like face, carved out of rock. And "massive" is no exaggeration. It is around three kilometers in length and is roughly half that distance wide. Clearly, if not just a trick of the light, the Face represents an incredible example of massive and radical alteration to the landscape, which, in all likelihood, would have taken place tens of thousands of years ago.

It was on July 25, 1976, that the Face on Mars caught the attention of NASA staff. That is when NASA's *Viking 1* spacecraft secured a series of aerial photos of Cydonia, including the now-legendary Face. That the pictures are now in the public domain has given rise to a great deal of thought on what, exactly, the Face represents. If it is not a natural formation, who constructed it and why? Let's see what Tonnies had to say about the Face on Mars, including the possibility that it represents one of the last pieces of evidence suggesting that Mars was once a bustling world, teeming with life.

Mac Tonnies's interest in the Face began in the 1980s, when he was still a teenager. It was more than a decade later that Tonnies decided to dig into the heart of the enigma, pursuing the truth of the mystery. In an extensive interview with me, Tonnies said:

"I've always had an innate interest in the prospect of extraterrestrial life. When I realized that there was an actual scientific inquiry regarding the Face and associated formations, I realized that this was a potential chance to lift SETI [the search for extraterrestrial intelligence] from the theoretical arena; it's within our ability to visit Mars in person. This was incredibly exciting, and it inspired an interest in Mars itself—its geological history, climate, et cetera. I have a B.A. in creative writing. So, of course, there are those who will happily disregard my book because I'm not 'qualified.' I suppose my question is, Who *is* qualified to address potential extraterrestrial artifacts? Certainly not NASA's Jet Propulsion Laboratory, whose Mars exploration timetable is entirely geology-driven."

But how, exactly, did the controversy begin, and what was it that led to so much interest—obsession, even—in Mars's massive mystery? Tonnies provides us with the timeline:

"NASA itself discovered the Face, and even showed it at a press conference, after it had been photographed by the Viking mission in the 1970s. Of course, it was written off as a curiosity. Scientific analysis would have to await independent researchers. The first two objects to attract attention were the Face and what has become known as the D&M Pyramid, both unearthed by digital imaging specialists Vincent DiPietro and Gregory Molenaar. Their research was published in *Unusual Martian Surface Features*; shortly after, Face researcher Richard Hoagland pointed out a collection of features—some, eerily pyramid-like—near the Face, which he termed the 'City.'"

As Tonnies noted, NASA—publicly, at least—dismissed the Face on Mars as nothing but a regular piece of rock that, superficially speaking, resembled a human face. And that was all—for NASA, if not for Tonnies. It's important to note Tonnies's words, which strongly suggest that the Face is not a natural creation. He said: "When NASA dismissed the Face as a 'trick of light,' they cited a second, discomfirming photo allegedly taken at a different sun-angle. This photo never existed. DiPietro and Molenaar had to dig through NASA archives to find a second image of the Face—and, far from disputing the face-like appearance, it strengthened the argument that the Face remained face-like from multiple viewing angles."

"The prevailing alternative to NASA's geological explanation is that we're seeing extremely ancient artificial structures built by an unknown civilization."

Tonnies notes the extent to which NASA went to try to discredit the theories suggesting that the Face on Mars was not a natural formation: "The prevailing alternative to NASA's geological explanation—that the Face and other formations are natural landforms—is that we're seeing extremely ancient artificial structures built by an unknown civilization. NASA chooses to ignore that there is a controversy, or at least a controversy in the scientific sense. Since making the Face public in the 1970s, NASA has made vague allusions to humans' ability to 'see faces' (e.g., the 'Man in the Moon') and has made lofty dismissals, but it has yet to launch any sort of methodical study of the objects under investigation. Collectively, NASA frowns on the whole endeavor. Mainstream SETI theorists are equally hostile."

TIME TRAVEL: THE SCIENCE AND SCIENCE FICTION

Tonnies made valuable observations relative to the controversy: "Basically, the Face—if artificial—doesn't fall into academically palatable models of how extraterrestrial intelligence will reveal itself, if it is in fact 'out there.' Searching for radio signals is well and good, but scanning the surface of a neighboring planet for signs of prior occupation is met with a very carefully cultivated institutionalized scorn. And of course it doesn't help that some of the proponents of the Face have indulged in more than a little baseless 'investigation.'"

With that all said, what did Mac Tonnies himself think of the Face on Mars? His thoughts, based upon years of dedicated research, make the answer to that question abundantly clear: "I think some of the objects in the Cydonia region of Mars are probably artificial. And I think the only way this controversy will end is to send a manned mission. The features under investigation are extremely old and warrant onsite archaeological analysis. We've learned—painfully—that images from orbiting satellites won't answer the fundamental questions raised by the Artificiality Hypothesis."

Tonnies suspected that the Face was a combination of a natural formation and something that had been radically altered into a new form. He elaborated on this particularly controversial line of thinking thus: "I suspect that we're seeing a fusion of natural geology and mega-scale engineering. For example, the Face is likely a modified natural mesa, not entirely unlike some rock sculptures on Earth but on a vastly larger and more technically challenging scale."

By making reference to "certain rock sculptures on Earth," Tonnies opened up a definitive can of worms, and it was practically impossible to close. It brings us

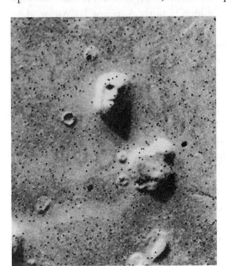

to the matter of Earth's own ancient history, specifically in Egypt. That the Cydonia region appeared to be the home of a couple of pyramid-like structures, as well as a Face that somewhat looked like the Sphinx of Giza, led Tonnies to note the following:

"There's a superficial similarity between some of the alleged pyramids in the vicinity of the Face and the better-known ones here on Earth. This has become the stuff of endless arcane theorizing, and I agree with esoteric researchers that some sort of link between intelligence on Mars and Earth deserves to be taken seriously. But the formations on Mars are much, much larger than terrestrial architecture. This suggests a significantly different purpose, assuming they're intelligently de-

Some people, including the late writer Mac Tonnies, have interpreted the geological "face" on Mars as purposeful construction and evidence that Mars was once a bustling world, teeming with life.

signed. Richard Hoagland, to my knowledge, was the first to propose that the features in Cydonia might be 'arcologies'—architectural ecologies—built to house a civilization that might have retreated underground for environmental reasons."

Tonnies was, as he admitted, highly concentrated on the fact that the carving was not just of a face but of what looked distinctly like a human face. Was it just a coincidence? Did the Martians resemble us? Was there a genetic connection between them and us? Yet again, Tonnies's thoughts and suspicions led him to formulate some incredible concepts:

"It's just possible that the complex in Cydonia—and potential edifices elsewhere on Mars—were constructed by indigenous Martians. Mars was once extremely Earth-like. We know it had liquid water. It's perfectly conceivable that a civilization arose on Mars and managed to build structures within our ability to investigate. Or the anomalies might be evidence of interstellar visitation—perhaps the remains of a colony of some sort. But why a humanoid face? That's the disquieting aspect of the whole inquiry; it suggests that the human race has something to do with Mars, that our history is woefully incomplete, that our understanding of biology and evolution might be in store for a violent upheaval. In retrospect, I regret not spending more time in the book addressing the possibility that the Face was built by a vanished terrestrial civilization that had achieved spaceflight. That was a tough notion to swallow, even as speculation, as it raises as many questions as it answers."

If the Face on Mars was built by intelligent beings—whether ancient humans or indigenous aliens—then how was such a monumental task achieved? On this point, we have the following from Tonnies, which is fairly restrained: "We need to bring archaeological tools to bear on this enigma. When that is done, we can begin reconstructing Martian history. Until we visit in person, all we can do is take better pictures and continue to speculate."

Working on the theory that Mars was once home to intelligent beings, we need to ask one of the most pressing questions of all: What was it that led the Martian world to become the dead world it is today? Tonnies says of this issue: "Astronomer Tom Van Flandern has proposed that Mars was once the moon of a tenth planet that literally exploded in the distant past. If so, then the explosion would have had severe effects on Mars, probably rendering it uninhabitable. That's one rather apocalyptic scenario. Another is that Mars's atmosphere was destroyed by the impact that produced the immense Hellas Basin [a 7,152-meter-deep basin located in Mars's southern hemisphere]. Both ideas are fairly heretical by current standards; mainstream planetary science is much more comfortable with Mars dying a slow, prolonged death. Pyrotechnic collisions simply aren't intellectually fashionable—despite evidence that such things are much more commonplace than we'd prefer."

So, what was it that prompted Tonnies to dig so deeply into the matter and write his acclaimed book *After the Martian Apocalypse*? Tonnies came straight to the point: "I was, frankly, fed up with bringing the subject of the Face on Mars up in online discussion and finding myself transformed into a straw man for self-professed experts. It was ludicrous. The book is a thought experiment, a mosaic of questions. We don't have all of the answers, but the answers are within our reach. Frustratingly, this has become very much an 'us vs. them' issue, and I blame both sides. The debunkers have ignored solid research that would undermine their assessment, and believers are typically quite pompous that NASA et al. are simply wrong or, worse, actively covering up."

Tonnies elected to comment further on this matter of a potential cover-up to hide an incredible revelation from the distant past. Commenting on NASA's Jet Propulsion Laboratory at the California Institute of Technology, he said: "When NASA/JPL released the first Mars Global Surveyor image of the Face in 1998, they chose to subject the image to a high-pass filter that made the Face look hopelessly vague. This was almost certainly done as a deliberate attempt to nullify public interest in a feature that the space agency is determined to ignore. So, yes, there is a cover-up of sorts. But it's in plain view for anyone who cares to look into the matter objectively. I could speculate endlessly on the forms a more nefarious cover-up might take—and I come pretty close in the book—but the fact remains that the Surveyor continues to return high-resolution images. Speculation and even some healthy paranoia are useful tools. But we need to stay within the bounds of verifiable fact lest we become the very conspiracy-mongering caricatures painted by the mainstream media."

It's important to note that Tonnies did not just focus his attention on the Face on Mars. He focused on other anomalies, too. One in particular fascinated Tonnies: "The Mars Global Surveyor has taken images of anomalous branching objects that look, for all the world, like organic phenomena. Arthur C. Clarke [the acclaimed science-fiction author, who died in 2008], for one, is sold on the prospect of large forms of life on Mars, and has been highly critical of JPL's silence. Clarke's most impressive candidates are what he has termed 'banyan trees' near the planet's south pole. And he collaborated with Mars researcher Greg Orme in a study of similar features NASA has termed 'black spiders'—root-like formations that suggest tenacious macroscopic life."

The Mars Global Surveyor, launched in 1996, "has taken images of anomalous branching objects that look, for all the world, like organic phenomena," according to writer Mac Tonnies.

In finality, Tonnies provided the following words: "Our attitudes toward the form extraterrestrial intelligence will take are painfully narrow. This is exciting intellectual territory, and too many of us have allowed ourselves to be told what to expect by an academically palatable elite. I find this massively frustrating. I hope *After the Martian Apocalypse* will loosen the conceptual restraints that have blinkered radio-based SETI by showing that the Face on Mars is more than collective delusion or wishful thinking. This is a perfectly valid scientific inquiry and demands to be treated as such."

Indeed, it does. Perhaps another effort at remote viewing is in order—a peek back through time to learn more about the origins of Mars's anomalous features.

CHAPTER 14

TIMELINES AND
THE ROSWELL INCIDENT

I s it possible that a theoretical time traveler could, as a result of his or her actions, provoke the spontaneous creation of a second timeline, maybe even a third or fourth? How about an unending number of timelines—a situation where there could be multiple Earths? Maybe, incredibly, millions? Welcome to the mind-boggling angle of what are known as timelines. *New Scientist* said in December 2019: "Would-be time travelers have long wrestled with the grandfather paradox: if you change things in the past and prevent yourself ever existing, how did you time travel in the first place? In other words, if Alice goes back in time and kills her grandfather Bob, she won't have been born and can't carry out her murderous plot. One way to avoid such paradoxes is the idea of branching universes, in which the universe we are in splits with each instance of time travel, creating two different universes."

"If you change things in the past and prevent yourself ever existing, how did you time travel in the first place?"

Also in 2019, Victor Tangermann of *Futurism* got into all of this head-scratching weirdness with an article titled "Paradox-Free Time Travel Possible with Many Parallel Universes." Tangermann wrote: "*New Scientist* reports that physicists Barak Shoshany and Jacob Hauser from the Perimeter Institute in Canada have come up with an apparent solution to these types of paradoxes that requires a very large—but not necessarily infinite—number of parallel universes.... 'The parallel universes approach that

One theory proposes that a time traveler could provoke the spontaneous creation of a new timeline or timelines—a situation where there could be multiple Earths.

we suggest says there are different parallel universes where things are roughly the same, and each one is mathematically on a separate space-time manifold,' Shoshany told *New Scientist.* 'You can go between those manifolds when you travel back in time.' …

"The model does have a major drawback, at least for narrative purposes: time travel won't do any good for your own timeline. 'What time travel means here is stepping between those histories—that's even freakier,' astrophysicist and dark matter expert Geraint Lewis at the University of Sydney, who was not involved in the research, told *New Scientist.* 'At some level it doesn't even feel like time travel anymore, because what's the point of going back and killing Hitler if the Second World War still plays out in the universe you're from?'"

> *"What time travel means here is stepping between those histories—that's even freakier."*

Now let's address the most famous UFO case of all, the Roswell incident of July 1947, and how it might relate to the phenomenon of multiple timelines.

Although there is an almost accepted assumption that Roswell involved either a spacecraft from another world or a Project Mogul balloon of the U.S. military, the fact is that there are more than a dozen explanations for what happened. I mention this to specifically demonstrate that, regardless of your personal opinion on the story, Roswell is nowhere near as resolved as many pro-extraterrestrial champions would have you believe, as you'll now see. Even more astounding, as a result of the timeline phenomenon, *all* of those theories for what happened at Roswell might actually be true. Prepare to learn about those amazing parallel worlds.

The late Jim Keith was the author of a number of books related to conspiracies and UFOs, including *Casebook on the Men in Black*, *The Octopus* (written with
Kenn Thomas), and *Black Helicopters over America*. In a small article titled "Roswell
UFO Bombshell," Keith described his clandestine meeting with "a longtime re-
searcher/instructor of engineering at a school in New Mexico" who claimed to know
the truth of Roswell. Keith wrote:

"According to my source, the true story behind the alleged UFO crash was
that there was an accident involving a B-29 flying from the Army Air Force Base in
Sandia (Albuquerque) to Roswell.... My source states that either an atomic bomb
or what is termed a 'bomb shape,' or 'test shape,' the shell of a nuke lacking explosives and atomic capability, and sometimes filled with concrete to add weight, was
accidentally or purposefully jettisoned above Corona, New Mexico, directly on the
flight path between Sandia and Roswell. Along with the bomb, metal foil used for
radar jamming, termed 'chaff,' may have also been dropped."

In 2010, Anomalist Books published the final title from the late Mac Tonnies:
The Cryptoterrestrials. Highly thought-provoking and deeply controversial in equal
measures, the book focused on the idea that UFOs are not the products of alien
races but of ancient terrestrial people that dwell deep underground and masquerade
as extraterrestrials to camouflage their true identity. Tonnies speculated that the Cryptoterrestrials are likely impoverished but utilize subterfuge, hologram-style technology, and staged events to suggest otherwise to us. He even theorized they may have
made use of large, balloon-style craft. On the matter of the Cryptoterrestrials using
balloons in covert missions, Tonnies said: "Maybe the Roswell device wasn't high
tech. It could indeed have been a balloon-borne surveillance device brought down
in a storm, but it doesn't logically follow that it was one of our own."

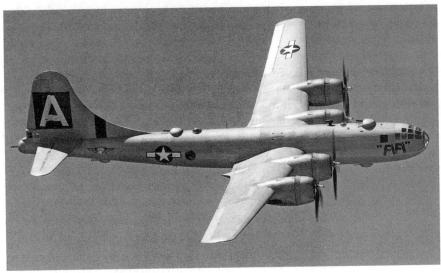

According to writer Jim Keith, "The true story behind the alleged UFO crash [at Roswell] was
that there was an accident involving a B-29."

TIME TRAVEL: THE SCIENCE AND SCIENCE FICTION

Annie Jacobsen's book *Area 51: An Uncensored History of America's Top Secret Military Base* created a huge wave of controversy when it was published in 2011, chiefly for one specific reason. The book includes a story suggesting that the Roswell craft and bodies were, in reality, the diabolical creations of a near-Faustian pact between the notorious "Angel of Death" Nazi Dr. Josef Mengele and Soviet premier Joseph Stalin.

The purpose of this early Cold War plan, said Jacobsen, was to plunge the United States into a kind of *War of the Worlds*-style panic by trying to convince the U.S. government that aliens were invading. And how would the plan work? By placing grossly deformed children (courtesy of the crazed Mengele) inside a futuristic-looking aircraft and then trying to convince the United States of the alien origins of both. Unfortunately for Stalin, we are told, the plot failed when a storm brought down the craft and its "crew" in the wilds of New Mexico. The event did not lead to widespread panic as planned but was hastily covered up by U.S. military authorities.

The Collins Elite is a quasi-official group within the U.S. government that believes the UFO mystery is one of demonic origins. Yes, forked tails, horns, fiery pits, and maybe even spinning heads and green vomit—those kinds of things. It must be stressed that the conclusions of the group were chiefly belief-driven rather than prompted by hard evidence.

One of the conclusions of the group was that Roswell was nothing less than a brilliant Trojan horse. For the members of the Collins Elite, deceptive demons had essentially used a kind of "cosmic alchemy" to create both the so-called memory metal found by Mack Brazel on the Foster Ranch and the curious bodies—or body parts—also located on the ranch. In other words, Roswell, as perceived by the Collins Elite, was an ingenious ruse, a staged crash, designed to have us humans believe that vulnerable extraterrestrials are in our midst when it's really the all-powerful minions of Satan. Hellfire!

Although no longer active in the UFO community, Timothy Cooper provoked a wealth of controversy in the 1990s thanks to an enormous body of allegedly leaked, and supposedly highly secret, documentation in his possession. They were documents which covered everything from crashed UFOs to alien autopsies and from sinister deaths in the UFO field to alien viruses. One such document—titled *UFO Reports and Classified Projects*—offers a non-UFO-themed explanation for what occurred at Roswell. The relevant extract reads as follows:

> One of the projects underway at that time incorporated re-entry vehicles containing radium and other radioactive materials combined with biological warfare agents developed by I.G. Farben for use against allied assault forces in Normandy in 1944. When a V-2 warhead impacted near the town of Corona, New Mexico, on July 4, 1947, the warhead did not explode and it and the deadly cargo lay exposed to the elements which forced the Armed Forces Special Weapons Project to close off the crash site and a cover story

was immediately put out that what was discovered was the remains of a radar tracking target suspended by balloons.

When personnel at the Roswell Army Airfield announced, in July 1947, that they had recovered a crashed flying disc, one thing was 100 percent absent: any mention of bodies. Needless to say, the body angle was also absent from the hasty follow-up explanation of a weather-balloon recovery. The body angle was also denied in the Air Force's July 1994 report on Roswell (titled "Report of Air Force Research Regarding the Roswell Incident"), as the following extract shows: "It should also be noted here that there was little mentioned in this report about the recovery of the so-called 'alien bodies.' The wreckage was from a Project Mogul balloon. There were no 'alien' passengers therein."

Three years later, however, things had changed. In a new document, "The Roswell Report: Case Closed," it was stated by the Air Force: "'Aliens' observed in the New Mexico desert were probably anthropomorphic test dummies that were carried aloft by U.S. Air Force high altitude balloons for scientific research.... The reports of military units that always seemed to arrive shortly after the crash of a flying saucer to retrieve the saucer and 'crew' were actually accurate descriptions of Air Force personnel engaged in anthropomorphic dummy recovery operations."

Then there is the matter of time travel. One of those who revealed his thoughts on this particular scenario was Lieutenant Colonel Philip Corso, coauthor with William Birnes of the alternately championed and denounced UFO-themed book *The Day after Roswell*. The unusual bodies found within the wreckage of the craft, Corso claimed, were genetically created beings designed to withstand the rigors of space flight, but they were not the actual creators of the UFO.

Corso gave much consideration to the idea that the Roswell UFO was a form of time machine, possibly even one designed and built by the denizens of Earth of a distant future rather than by the people of a faraway solar system.

Right up until the time of his death in 1998, Corso speculated on the distinct possibility that the U.S. government might still have no real idea of who constructed the craft or engineered the bodies found aboard. Notably, Corso gave much consideration to the idea that the Roswell UFO was a form of time machine, possibly even one designed and built by the denizens of Earth of a distant future rather than by the people of a faraway solar system.

Following the defeat of Nazi Germany in 1945, numerous German scientists were brought to the United States via a program called Operation Paperclip. Many of those scientists went to work at the New Mexico–based White Sands Missile Range (or the White Sands Proving Ground, as it was known back then). This has

TIME TRAVEL: THE SCIENCE AND SCIENCE FICTION

An alien exhibit at the International UFO Museum and Research Center, Roswell, New Mexico. Among the theories behind the infamous 1947 UFO incident near Roswell are that the craft was a time machine from future earthlings, a Nazi aircraft, a staged event meant to scare the Soviets, or a weather balloon.

led to a scenario involving a highly secret program to test-fly radical Nazi aircraft captured after Hitler and his cronies went belly-up.

Someone who has deeply addressed the Nazi/Roswell links is Joseph P. Farrell, the author of *Roswell and the Reich: The Nazi Connection*. Farrell, however, takes things in a decidedly different direction. He concludes the Roswell event was not caused by the flight and crash of a German craft flown from White Sands. Rather, Farrell believes that "what crashed may have been representative of an independent postwar Nazi power—an extraterritorial Reich monitoring its old enemy, America."

The Flying Saucer was a science-fiction novel published in 1948 and written by Bernard Newman, a man who penned more than a hundred books on subjects from real-life espionage to global politics and current affairs. It was his foray into the weird world of crashed UFOs that was perhaps most notable of all, however.

The book was published just 11 months after the alleged recovery by the U.S. military of a flying saucer in Lincoln County, New Mexico, in July 1947. *The Flying Saucer* tells the story of a secret cabal that stages a series of hoaxed UFO crashes with the express purpose of attempting to unite the world against a deadly alien foe that, in reality, does not exist.

That Newman had numerous high-level connections to officialdom has given rise to the theory that, first, Roswell itself was a staged event—one designed to scare the Soviets into thinking the United States had acquired alien technology, and, second, Newman based his novel on data secured by sources with top-secret knowledge of the Roswell ruse.

In an article titled "Roswell Explained—Again," published in *Fate* magazine in September 2005, Kevin Randle stated that during the early part of the 1990s, he "interviewed a man who worked with NASA at the White Sands Missile Range." The man in question was Gerald Brown.

According to Randle, Brown speculated that "some kind of flying wing had crashed while carrying five chimps dressed in silver flying suits." So far, no evidence has surfaced to suggest that the alien body stories can be explained away via the chimpanzee scenario. On the other hand, there is no hard, undeniable evidence for *any* theory when it comes to Roswell.

On July 28, 1995, a report surfaced from the National Security and International Affairs Division of the General Accounting Office (today called the Government Accountability Office) that disclosed the results of its investigation of the Roswell affair. Commenting on an Air Force report on Roswell published in July 1994, the GAO noted the following:

> DOD informed us that the U.S. Air Force report of July 1994, entitled Report of Air Force Research Regarding the Roswell Incident, represents the extent of DOD records or information concerning the Roswell crash. The Air Force report concluded that there was no dispute that something happened near Roswell in July 1947 and that all available official materials indicated the most likely source of the wreckage recovered was one of the project MOGUL balloon trains. At the time of the Roswell crash, project MOGUL was a highly classified U.S. effort to determine the state of Soviet nuclear weapons research using balloons that carried radar reflectors and acoustic sensors.

And that's where things stand to this day. The theories range from the plausible to the unlikely, with some hazily hovering somewhere in between. A few of those scenarios may be the work of hoaxers or "Walter Mitty" types. Others—probably most of them—are almost certainly born out of the clandestine worlds of disinformation and psychological warfare. Somewhere in this confusing mass (and mess) of theories, the truth of Roswell exists. Or, to recall the issue of timelines, maybe *each and every single scenario is real*—albeit in a very weird way. Consider this: Some witnesses to the crash were sure they saw crash-test dummies. Others have said they saw handicapped people used in high-altitude aircraft. And, of course, many have sworn they came across the bodies of extraterrestrials.

If the Roswell craft was a time machine, perhaps its crash in the wilds of New Mexico provoked the sudden, spontaneous creation of a dozen situations, all real but wholly different.

TIME TRAVEL: THE SCIENCE AND SCIENCE FICTION

It's not at all impossible that if the Roswell craft was a time machine, perhaps its crash in the wilds of New Mexico provoked the sudden, spontaneous creation of a dozen situations, all real but wholly different. Theoretically, a chaotic situation that involves the witnesses seeing brief glimpses of parallel worlds—as timelines mesh and spread like ripples on a pond—may not be as implausible as it seems. Such a situation just might explain why there can be so many very different scenarios for what took place at Roswell all those years ago.

Now let's take a look at the U.K.'s most famous UFO event—an event that, just like the Roswell affair, may have had its origins in the world of time travel.

In the final days of December 1980, multiple strange encounters and wild incidents occurred in Rendlesham Forest, Suffolk, England, across a period of three nights. Based on their personal encounters, many of those who were present believed that something almost unbelievable came down in the near-pitch-black woods on

A series of strange encounters and wild incidents in Rendlesham Forest, Suffolk, England, in December 1980 may have involved, according to one witness, visitors from a far-flung future.

the night of December 26. Lives were altered forever—and for the most part not for the better, I need to stress. Many of those who were present on those fantastic nights found their minds dazzled, tossed, and turned—and incredibly quickly, too.

Those incidents involved American military personnel who, at the time it all happened, were stationed in the United Kingdom. Their primary role was to provide significant support in the event that the Soviets decided to flex their muscles just a bit too much—or, worse still, planned on hitting the proverbial red button and ending civilization in hours or even minutes.

To understand exactly what occurred in the woods on those fateful nights, it's important that we go back to the beginning. We can examine the events through a now-legendary memo on the incidents that was carefully prepared by Lieutenant Colonel Charles Halt. At the time, he was the deputy base commander at RAF Bentwaters, a Royal Air Force base (now a commercial property called Bentwaters Parks) near Woodbridge, Suffolk. On January 13, 1981, Halt prepared the following report of a sighting at the nearby RAF Woodbridge; it was sent to the U.K. Ministry of Defense (MoD) for scrutiny. It gives a fairly brief, but certainly detailed, account of what happened:

1. Early in the morning of 27 Dec 80 (approximately 0300L) two USAF security police patrolmen saw unusual lights outside the back gate at RAF Woodbridge. Thinking an aircraft might have crashed or been forced down, they called for permission to go outside the gate to investigate. The on-duty flight chief responded and allowed three patrolmen to proceed on foot. The individuals reported seeing a strange glowing object in the forest. The object was described as being metallic in appearance and triangular in shape, approximately two to three meters across the base and approximately two meters high. It illuminated the entire forest with a white light. The object itself had a pulsing red light on top and a bank(s) of blue lights underneath. The object was hovering or on legs. As the patrolmen approached the object, it maneuvered through the trees and disappeared. At this time the animals on a nearby farm went into a frenzy. The object was briefly sighted approximately an hour later near the back gate.

2. The next day, three depressions 1.5 inches deep and 7 inches in diameter were found where the object had been sighted on the ground. The following night (29 Dec 80) the area was checked for radiation. Beta/gamma readings of 0.1 milliroentgens were recorded with peak readings in the three depressions and near the center of the triangle formed by the depressions. A nearby tree had moderate (0.05–0.07) readings on the side of the tree toward the depressions.

3. Later in the night a red sun-like light was seen through the trees. It moved about and pulsed. At one point it appeared to throw off glowing particles and then broke into five separate white objects and then disappeared.

TIME TRAVEL: THE SCIENCE AND SCIENCE FICTION

Immediately thereafter, three star-like objects were noticed in the sky, two objects to the north and one to the south, all of which were about 10 degrees off the horizon. The objects moved rapidly in sharp, angular movements and displayed red, green and blue lights. The objects to the north appeared to be elliptical through an 8–12 power lens. They then turned to full circles. The objects to the north remained in the sky for an hour or more. The object to the south was visible for two or three hours and beamed down a stream of light from time to time. Numerous individuals, including the undersigned, witnessed the activities in paragraphs 2 and 3.

Charles I. Halt, Lt Col, USAF
Deputy Base Commander

One of the key figures in the Rendlesham affair was Jim Penniston. He entered the U.S. Air Force in 1973. At the time all hell broke loose in Rendlesham Forest, Penniston was a senior security officer. Both he and Halt had startling encounters on those December nights. Penniston actually touched the whatever-it-was, an act that he now believes caused him to receive a binary code message that was essentially downloaded into his mind.

Techopedia explains what, precisely, binary codes are: "Binary code is the most simplistic form of data. It is represented entirely by a binary system of digits consisting of a string of consecutive zeros and ones. Binary code is often associated with machine code in that binary sets can be combined to form raw code, which is interpreted by a computer or other piece of hardware."

While under hypnosis, Penniston stated that the presumed aliens are, in reality, visitors from a far-flung future. Our future.

Formerly of the U.S. Air Force and one of the key military players in the famous UFO encounter at Rendlesham Forest in December 1980, Sergeant Penniston eventually, in 1994, underwent hypnotic regression in an attempt to recall deeply buried data relative to what occurred to him during one of Britain's closest encounters. Interestingly, while under hypnosis, Penniston stated that the presumed aliens are, in reality, visitors from a far-flung future. Our future. That future, Penniston added, is very dark, in infinitely deep trouble, and polluted. In it, the human race is overwhelmingly blighted by reproductive problems. The answer to those massive problems, Penniston was told by the entities he met in the woods, was for the entities to travel into the distant past—to our present day—to secure sperm, eggs, and chromosomes in an effort to ensure the continuation of the severely waning human race of tomorrow.

Ian Ridpath, a skeptic of the UFO story, said: "In the days following the events of December 26 and 28, Halt and Penniston were debriefed separately by [Colonel]

TIME TRAVEL: THE SCIENCE AND SCIENCE FICTION

Ted Conrad, Halt's boss. These were the only two witnesses Conrad talked to. It is sometimes claimed in UFO circles that the witnesses were subjected to interrogation by the Air Force Office of Special Investigations (AFOSI), including injections by the so-called 'truth drug' sodium pentothal." In January 2010, however, Conrad denied that drugs were used.

UFO researcher Dr. David Clarke was told by Conrad: "There were no conspiracies, no secret operation, no missile accident, and no harsh interrogations by OSI. I was in a position to know about the OSI. They had their own chain of command, but in practice the OSI commander kept me informed of any ongoing investigations they had."

Try telling all of that to the personnel in the woods on those cold nights in December 1980.

CHAPTER 15

LITTLE GREEN TIME SURFERS

ithin the field of ufology, the term "little green men" turns up with reg-
ularity, usually in a disparaging, fun-poking, debunking fashion. But despite its discredited place in popular culture, it's a fact that reports of little
green men are not uncommon in UFO research. Certainly, one of the most infamous
examples can be found in a certain saga that occurred on the night of August 21, 1955.

The location was a farmhouse situated between the towns of Kelly and Hopkinsville, Kentucky. What began with sightings of strange lights in the sky soon expanded into a definitive shoot-out with a group of little, green-colored men. The
farmhouse belonged to the Sutton family, who wasted no time in trying to blast the
diminutive things into oblivion. It's a case that remains a ufological favorite. The association between aliens and small, green humanoids isn't exclusively tied to the
Kelly–Hopkinsville case, however.

A researcher of folklore named Chris Aubeck has been able to determine that
those three now-famous words—"little green men"—date back to the final years of
the nineteenth century. In 1899, for example, the *Atlanta Constitution* published a
story written by Charles Battell Loomis titled "Green Boy from Hurrah"—the "hurrah" of the title being a reference to another world. The renowned fantasy novelist
Edgar Rice Burroughs wrote of the "green men of Mars" in his 1906 story "A Prin-
cess of Mars."

On May 22, 1947, just one month before the term "flying saucer" was coined
(following the sighting of a squadron of UFOs seen over Mount Rainier, Washing-

The term "little green men" as a reference to space aliens—perhaps time-traveling ones—dates back to the final years of the nineteenth century.

ton, by a pilot named Kenneth Arnold), a reference to "little green men" was made in a Harrisburg, Illinois, newspaper, the *Daily Register*. Although the small article was on a medical condition that can cause the skin to turn green, it demonstrates that the terminology was in popular use before UFOs were a part of our culture. Four years later, in 1951, Mack Reynolds's sci-fi novel *The Case of the Little Green Men* was published. The subject matter was alien entities infiltrating the human race.

We may never know for sure exactly when the "LGM" term was created; however, when it comes to the matter of potential ancient aliens, we find a link between the phenomenon and those now widely used three words. It's a story that takes us to the east coast of England and centuries back in time. And, by that, I mean literally back in time.

So the tale goes, back in the twelfth century, a young girl and boy of strangely green-hued skin appeared in the English village of Woolpit, which is situated between the towns of Bury St. Edmunds and Stowmarket. The children claimed to have come from a magical place called St. Martin's Land, which existed in an atmosphere of permanent twilight and where the people lived underground on nothing but green beans. While the story has been relegated by many to the realms of mere myth and folklore, it might not be just that. It might actually be much more.

> *While the story has been relegated by many to the realms of mere myth and folklore, it might not be just that. It might actually be much more.*

According to the old legend, the two children remained in Woolpit and were ultimately baptized by the villagers, who accepted them as their very own. And although the boy grew sickly and eventually died, the girl did not. She thrived and finally lost her green-tinged skin to acquire the normal-colored skin of healthy appearance. According to the legend, she also, somewhat in defiance of the disapproving attitudes of certain members of the village, became, as it was amusingly termed back then, "rather loose and wanton in her conduct."

The story shows no signs of going away anytime soon. In fact, the people of the village of Woolpit embrace the story to this very day, as the following extract from the village's official website demonstrates:

"In 1016 Ulfketel, Earl of East Anglia, granted the church and manor of Wlfpeta to the Abbey of St Edmunds. The monks received ten marks yearly from this grant, but the King appropriated the revenues for the benefit of one of his officials. A monk named Sampson determined to put matters right and in 1159 traveled to Rome to obtain a charter from Pope Alexander III. Despite being captured and robbed by his enemies in the course of his journey, Sampson managed to preserve the Pope's precious letter directing the reversion of Woolpit and its church to the monks, and returned to England after three years. The monks were once again able to enjoy the income, and Sampson was later appointed Abbot.

"One prominent feature of the village sign is two small children. They depict a story that goes back to the 12th century and tells the legend of 'The Green Children of Woolpit'. This curious tale is recorded as taking place at about the same time as Sampson's journey to Rome. Very briefly, reapers were astonished at the discovery of a boy and a girl in a hole in the ground. The children were green, and spoke no recognizable language. The boy, who was sickly, soon died, but the girl grew up in Woolpit, and is said to have married a man from King's Lynn. This story has been re-enacted on many occasions and has appeared on children's television."

Ralph of Coggeshall, a thirteenth-century monk, wrote specifically of the strange girl, who fortunately survived and thrived in Woolpit all those centuries ago: "Being frequently asked about the people of her country, she asserted that the inhabitants, and all they had in that country, were of a green color; and that they saw no sun, but enjoyed a degree of light like what is after sunset.

"Being asked how she came into this country with the aforesaid boy, she replied, that as they were following their flocks, they came to a certain cavern, on entering which they heard a delightful sound of bells; ravished by whose sweetness, they went for a long time wandering on through the cavern, until they came to its mouth.

"When they came out of it, they were struck senseless by the excessive light of the sun, and the unusual temperature of the air; and they thus lay for a long time. Being terrified by the noise of those who came on them, they wished to fly, but they could not find the entrance of the cavern before they were caught."

That both wild children were reportedly green-skinned and lived underground in a mysterious locale has led many to disregard the tale out of hand as one of fairy-based, mythological proportions and nothing else whatsoever. That may not actually have been the case, however.

The village sign for Woolpit, Suffolk, depicts the fabled green children of the twelfth century. According to one theory, the children were neither local feral children nor aliens but evolved human visitors from the future.

The pair may have been suffering from a condition called hypochromic anemia, in which the sufferer, as a result of a very poor diet that, in part, affects the color of the red blood cells, can develop skin of a noticeably green shade. In support of this scenario, hypochromic anemia was once known as chlorosis, a word formulated in the early 1600s by a Montpellier professor of medicine named Jean Varandal. And why did Varandal choose such a name? Simple: it came from the Greek word *chloris*, meaning greenish-yellow or pale green.

Therefore, we might well conclude that the strange children of Woolpit were definitively wild in nature. And given their state of poor health, they may certainly have lived poor and strange lives, whether underground or permanently under a thick canopy of dense forest of the type that dominated England at that time, with others of their kind, in and around Suffolk, just as they had claimed to the villagers of Woolpit. And they may have all the while struggled to survive on the meager supplies of food available to them, which were ultimately responsible for their greenish-yellow hue.

There is another possibility, however: these little green children may not have been of this world. They may have been ancient extraterrestrials.

TIME TRAVEL: THE SCIENCE AND SCIENCE FICTION

Dr. Karl Shuker, an authority on the mysteries of our world and someone who has spent a great deal of time investigating the saga of the green children of Woolpit, says: "A very different and far more dramatic explanation was proffered by Harold T. Wilkins, an investigator of unexplained anomalies. In his book *Mysteries: Solved and Unsolved* (1959), Wilkins boldly proposed that the green children may have entered our world from a parallel version (existing in a separate dimensional plane but directly alongside our own), by accidentally passing through some form of interdimensional 'window' bridging the two. Another dramatic proposal is that the green children are extraterrestrials. As long ago as 1651, Robert Burton opined in his tome *Anatomy of Melancholy* that they may have come from Venus or Mars."

On a similar path, blogger Tristan Eldritch notes: "To a modern reader, the children might suggest stranded extraterrestrials of some kind, and this interpretation of the tale goes back much further than you would imagine." Indeed, as Eldritch also reveals, "the story found its way into Francis Goodwin's fancy *The Man in the Moone* (1638), a work sometimes regarded as the very earliest example of science fiction."

Researcher Melanie Koslovic says: "It has also been suggested that the children may have been aliens, or inhabitants of a world beneath the Earth. In 1996 an article was published in the magazine *Analog* concerning this mystery; astronomer Duncan Lunan hypothesized that the children were accidentally sent to Woolpit from their home planet due to a malfunction in their 'matter transmitter.'

"Lunan thinks that the planet from which the children originated may be trapped in a synchronous orbit around its sun, presenting the conditions for life only in a narrow twilight zone between a fiercely hot surface and a frozen dark side. He believes the children's green color is a side effect of consuming genetically modified alien plants eaten on their planet."

A writer for *HubPages* notes of Lunan's theory, in an article titled "Green Children of Woolpit: Mysterious Visitors from an Unknown Land": "Encouraged by the introduction of this element of science fiction to the proceedings, another unusual hypothesis has surfaced. The foundation of this theory is based on the belief that the children came from another dimension. Proponents of this reasoning believe that the pair inadvertently discovered a cave opening which was a portal between their dimension, and our own. Evidently, the portal was not a two-way access point; or once the children were through, it vanished: as no one was ever able to pinpoint such an opening near the wolf pit where the two were discovered."

Duncan Lunan himself says that the green girl ultimately became the wife of one "Richard Barre, one of Henry II's senior ambassadors, which rather puts paid to the 'runaways from some primitive tribe' class of explanations. Her first, illegitimate child may have been fathered by Henry himself. I've traced her descendants to the present—one of them was deputy head of the House of Lords under Margaret Thatcher, and he thinks it's a hoot: 'I knew my ancestors were colorful, but not that colorful.' It looks as if the children grew up in a human colony on a planet with a trapped rotation and were returned to Earth in a matter-transmitter accident, one of a number which happened while the Earth's magnetic field was disturbed by the most violent solar activity since the Bronze Age."

Before you write off Lunan's theory as being far too fanciful, it's worth noting that he attended Marr College and Glasgow University. He is an M.A. with honors in English and philosophy, with physics, astronomy, and French as supporting subjects, and has a postgraduate diploma in education.

Such is the age of the curious tale of Woolpit's centuries-old saga of the village's eerie, green-skinned children that we'll likely never know if they really were ancient aliens or of far more down-to-earth origins. Whatever the truth, of one thing we can be sure: the enchanting nature of the story ensures that, with every new generation that comes along, it gathers more momentum and provokes increased fascination.

Paranormal writer Brent Swancer says: "The enigmatic children have been alternately theorized as being from a parallel dimension, aliens, inhabitants of some hidden subterranean civilization, feral children raised by animals, orphans with arsenic poisoning or malnutrition which made their skin green, or Flemish immigrant refugees fleeing persecution and a battle that had killed their parents. In the end, no one really knows who they were, where they came from, or indeed how much of the story is even based in reality, and so the story of the Green Children of Woolpit has become an enduring mystery that will perhaps never be solved."

"They came from a time in Earth's future, when climate change has caused the planet's atmospheric conditions to degrade so drastically that no one can see the sun anymore and everyone's turned green."

Well, maybe the mystery has been solved. Now it's time for me to share with you fascinating material that may be evidence of time travelers.

At the *Time Travel Nexus*, C. R. Berry definitely gives us something to think about. Berry notes the extraterrestrial theory but adds that some "have suggested that the children were not aliens but humans from the future." Berry expands:

"They walked through a time slip and ended up in the 12th century, similar to what happened to Charlotte Anne Moberly and Eleanor Jourdain. They came from a time in Earth's future, when climate change has caused the planet's atmospheric conditions to degrade so drastically that no one can see the sun anymore and everyone's turned green. As Sir David Attenborough predicted very recently, it's led to

TIME TRAVEL: THE SCIENCE AND SCIENCE FICTION

the collapse of civilisation as we know it and the extinction of 'much of the natural world.' It's left pockets of survivors, the remnants of humanity, trying to make their way in a new, bleak world, and religion has come to the fore again. 'St. Martin's Land' is one of these new settlements. This would explain their unknown language and unfamiliar clothes. Their language could've been Modern English, or perhaps a future form of English—both of which would've been incomprehensible to the Early Middle English speakers of the 12th century. If true, the Green Children of Woolpit are some of the earliest TTTs we have on record."

CHAPTER 16

THE MYSTERY OF THE MISSING "THUNDERBIRD PHOTO"

I t's time to share with you a saga that makes it abundantly clear just how mind-boggling the phenomenon of time travel really is. You are about to come face-to-face with the bizarre matter of a certain vanished photograph that, if it could be found, just might provide evidence that time can indeed be manipulated to incredible degrees. It concerns a legendary creature that is a significant part of Native American lore and legend: nothing less than a huge, winged creature known as the Thunderbird. Before we get to the story of time travel, however, let's have a look at the phenomenon of the Thunderbird.

The website *Legends of America* says: "Described as a supernatural being, the enormous bird was a symbol of power and strength that protected humans from evil spirits. It was called the Thunderbird because the flapping of its powerful wings sounded like thunder, and lightning would shoot out of its eyes. The Thunderbirds brought rain and storms, which could be good or bad—good when the rain was needed, or bad when the rain came with destructive strong winds, floods, and fires caused by lightning. The bird was said to be so large that several legends tell it picking up a whale in its talons. They were said to have bright and colorful feathers, with sharp teeth and claws. They were said to live in the clouds high above the tallest mountains. Various tribes have different oral traditions about the magical Thunderbird, which they both highly respected and feared. In Gros Ventre traditions, it was the Thunderbird who gave the sacred pipe to the people. Some Plains tribes, includ-

The Thunderbird was a figure in many Native American communities, including the Gros Ventre, the Arapaho, the Kwakiutl, and the Ho-Chunk. This Thunderbird totem appears in Thunderbird Park, Victoria, British Columbia, Canada.

ing the Arapaho, associated Thunderbirds with the summer season, while White Owl represented the winter season."

There is this from *Native Languages*, too: "The Thunderbird is a widespread figure in Native American mythology, particularly among Midwestern, Plains, and Northwest Coast tribes. Thunderbird is described as an enormous bird (according to many Northwestern tribes, large enough to carry a killer whale in its talons as an eagle carries a fish) who is responsible for the sound of thunder (and in some cases lightning as well). Different Native American communities had different traditions regarding the Thunderbird. In some tribes, Thunderbirds are considered extremely sacred forces of nature, while in others, they are treated like powerful but otherwise ordinary members of the animal kingdom. In Gros Ventre tradition, it was Thunderbird (Bha'a) who gave the sacred pipe to the people. Some Plains tribes associated thunderbirds with the summer season (in Arapaho mythology, Thunderbird was the opposing force to White Owl, who represented winter). Thunderbirds are also used as clan animals in some Native American cultures. Tribes with Thunderbird Clans include the Kwakiutl and Ho-Chunk tribes. On the Northwest Coast, the thunderbird symbol is often used as a totem pole crest."

The website of the Cuyamungue Institute in New Mexico provides data, too: "The wingspan of the Thunderbird was described to be twice as long as a Native Indian war canoe. Underneath its wings are lightning snakes which the Thunderbird uses as weapons. Lightning is created when the Thunderbird throws these lighting snakes or when he blinks his eyes that glow like fire. Sometimes these lightning snakes are depicted in Native American art as having wolf or dog-like heads with serpent tongues. They are occasionally referred to as the Thunderbird's dogs. Native American art portrays the Thunderbird with a huge curving beak and prominent ears or horns. The Thunderbird is large and strong enough to hunt its favorite food which is the killer whale. The lightning snakes of the Thunderbird are used during hunts out at sea for the killer whale. After capture, the Thunderbird carries the killer whale back to the mountain to eat. According to legend, the Thunderbird and killer whale once battled so hard that entire trees were uprooted. This was the explanation why there are treeless prairie regions near the Pacific Northwest Coast mountains."

Now let's get to the crux of the affair. It's the matter of a vanished photograph of just such a creature that numerous people in the fields of cryptozoology and the paranormal swear they have seen but that, bafflingly, can no longer be found. Anywhere. It's as if the photo has been erased from our timeline—something, you may recall, that almost happened to Marty McFly in the iconic movie *Back to the Future*. That may be precisely what happened in the real world. Notably, while portions of the timeline are gone, the memories and recollections of those who saw the Thunderbird picture before it vanished remain. Like the events at Roswell, New Mexico, in 1947, and at Rendlesham Forest, England, in 1980, memories remain, even though reality and timelines are altered and parallel worlds might be created as a strange consequence.

> *Notably, while portions of the timeline are gone, the memories and recollections of those who saw the Thunderbird picture before it vanished remain.*

To see how things really began, and in relation to the matter of this elusive photograph, we have to turn our attention to the folks at *Mothman Fandom*: "In the May 1963 issue of *Saga* magazine, writer Jack Pearl recounted this story of the Tombstone Thunderbird, along with some large bird sightings of the early 1960s. Not only did he tell the story though, he went one step further and claimed that the *Tombstone Epitaph* had, in 1886, 'published a photograph of a huge bird nailed to a wall. The newspaper said that it had been shot by two prospectors and hauled into town by wagon. Lined up in front of the bird were six grown men with their arms outstretched, fingertip to fingertip. The creature measured about 36 feet from wingtip to wingtip.'"

The story grows and grows. Take note of the words of the staff of *Truewest Magazine*: "Another writer, H. M. Cranmer, contended in *Fate* magazine in the fall of 1963 that the picture had been published in newspapers all over the country. Ivan T.

TIME TRAVEL: THE SCIENCE AND SCIENCE FICTION

Sanderson, considered an eminent researcher in the study of strange phenomena, claimed to not only seeing the photo, but once having a photocopy he unfortunately loaned out and never got back. Someone later came forward and remembered seeing Sanderson display the photo on Canadian television, although no copies of the show have been found."

Dr. Karl Shuker, one of the leading figures in the field of cryptozoology, has taken keen note of the story of the Thunderbird and that weirdly vanished image. He said that one of the things he did was to chase down "a copy of the thunderbird photo [that] was displayed on television by American cryptozoologist Ivan T. Sanderson during the early 1970s, when appearing as a guest in an episode of the long-running Canadian series *The Pierre Berton Show*." Shuker continues:

"Consequently, I contacted the Audio-Visual Public Service division within the National Archives of Canada, to enquire whether a copy of the Sanderson episode in this series had been preserved. Unfortunately, however, I learnt from research assistant Caroline Forcier Holloway that she had been unable to locate this particular episode, and needed a precise production or release date for it in order to continue looking, because there were 597 episodes in this series still in existence, each of which contained more than one guest. Moreover, there were others that seemed to have been lost, so there was no guarantee that the episode containing Sanderson was among the 597 preserved ones anyway. However, one of my correspondents, Prof. Terry Matheson, an English professor at Saskatchewan University with a longstanding interest in the thunderbird photo, claimed in a letter to me of 22 September 1998 that Sanderson appeared on *The Pierre Berton Show* not in the early 1970s, but actually no later than the mid-1960s."

Troy Taylor at *Weird U.S.* has a say in all this, too: "So, is the photo real? If not, then why do so many of us (myself included) with an interest in the unusual claim to

Many witnesses claim to have seen a rumored photograph of men posing with a gigantic bird, but the photo itself has never been found.

remember seeing it? Who knows? Just recently, in the late 1990s, author John Keel insisted, 'I know I saw it! And not only that—I compared notes with a lot of other people who saw it." Like many of us, Keel believes that he saw it in one of the men's magazines (like *Saga* or *True*) that were so popular in the 1960s. Most of these magazines dealt with amazing subject matter like Bigfoot and ghosts. Keel also remembers the photo in the same way that most of us do—with men wearing cowboy clothing and the bird looking like a pterodactyl or some prehistoric, winged creature."

In the early 2000s, I had some significant correspondence with the late anomalies researcher—and the editor and publisher of *Strange Magazine*—Mark Chorvinsky, who died in 2005. He sent me the following that was to be used in a full-length book on the Thunderbird photograph that I ultimately shelved. Nevertheless, Chorvinsky was good enough to allow me to use the material should I one day decide to return to the mystery of the Thunderbird. I will provide his words without interruption:

"The legend of the Thunderbird Photo is intertwined with an article that is said to have appeared in the April 26, 1890 *Tombstone Epitaph*. Legend has it that the photograph accompanied the article. Until now there was a question as to whether or not any Thunderbird article appeared in the *Epitaph*. Despite the fact that the *Epitaph* article had been reprinted several times in the last century, local Tombstone historians as well as the editor of a later version of the *Epitaph* claimed that the article never ran. To a large extent this says more about the incompetence of those who have supposedly researched this case than anything else.

"I interviewed Tombstone historian Ben Traywick who told me that he had searched in vain through the entire run of the *Tombstone Epitaph* for the Thunderbird article. 'There was no article on the Thunderbird,' Traywick claimed. 'It may have appeared in another paper, but not the *Epitaph*.' Traywick also makes this statement in our *Strange World Video* (*Strange Magazine*, 1995, dir. Mark Chorvinsky/Greg Snook).

"Additionally, Wallace E. Clayton, editor of *The National Tombstone Epitaph,* wrote in the November 1984 issue in response to a reader's question about the Thunderbird Photograph that *Epitaph* contributor Ben T. Traywick spent a great amount of time in libraries in Arizona and the major Western history research institutions in California, and never has found the photograph nor any mention of the big bird in the newspapers of the time.

"I interviewed Ben Traywick again recently and he reiterated that he had searched local papers for the years in question, and had never seen the *Epitaph* article. But, he explained, Tombstone resident Jack Fiske had claimed to have seen the article, although he never showed it to Traywick. I then called Jack Fiske, who I had also interviewed earlier in my investigation, and asked about the article.

TIME TRAVEL: THE SCIENCE AND SCIENCE FICTION

"'I never saw it,' Fiske told me. 'You should talk to Ben Traywick, he has the article.' The Tombstone 'experts' clearly did not have the article. Richard Ravalli, Jr., in the 'Letters' section of *Strange Magazine* #18 (Summer, 1997) wondered about Ben Traywick's statement that there was never a T-bird article—with or without a photo—in the *Tombstone Epitaph*. Ravalli noted that the only reprinting of the article that he was able to find was in Horace Bell's *On the Old West Coast*. Ravalli asked, 'My question is this: Does the article exist? Or did Bell (or someone else) make it up?'

"With Tombstone locals Traywick and Fiske—both of whom had written articles about the Tombstone monster—unable to come up with an article, and the only readily available source being Bell's reprinting in his book, confusion on this issue has reigned for a number of years. I decided to determine definitively whether or not the article had ever been in the *Epitaph*. The date given for its original publication was April 26, 1890. According to the Library of Congress, the only library known to have a full run of the 1890 *Tombstone Epitaph* is the University of Arizona. I deputized University of Arizona reference librarian Jodi Nuñez and her capable crew. They were willing and able to check out their newspaper collection and in a short period of time they confirmed that the Thunderbird article was in fact in the *Epitaph* and within several days I was able obtain a copy of the original *Epitaph* article.

"Traywick and Clayton are clearly incorrect about the article never appearing in the *Epitaph*—they had only to search that newspaper on the date that the article was said to have appeared to have found it. Likewise John A. Keel, Mark Hall, Ivan T. Sanderson, and other investigators who have supposedly investigated this case.

"In his *Monsters, Giants, and Little Men from Mars* (Dell, 1975: NY, p. 175), strange phenomena popularizer Daniel Cohen perpetuated the notion that the Tombstone flying monster article may have never appeared in the *Epitaph,* writing that the story '… was said to have first seen the light of day in the *Tombstone Epitaph,* though, again, no one seems able to locate the original account. The story, however, has been retold again and again in articles and columns about Western oddities, and in Fortean publications, and there is a considerable variation on the versions now extant.'

"In *The Encyclopedia of Monsters* (Avon, NY, 1991, p. 85—first published 1982), Daniel Cohen states flatly that the Tombstone flying monster story was 'attributed (incorrectly) to the *Tombstone Epitaph*.

"This is my first opportunity to correct Cohen, Traywick, and others' claims that the article was either never published in the *Epitaph* or was somehow 'unfindable,' which has added a certain amount of confusion to an already complicated case. In any event, we now know that the article appeared in the April 26, 1890 *Tombstone Epitaph*.… The article is fascinating, but there is no photo accompanying it, nor does the article mention a photograph.

"According to the article, two ranchers were riding in the desert between Whetstone, Arizona, and the Huachuca Mountains in late April, 1890, when they came upon a 'winged monster resembling a huge alligator with an extremely elongated tail and an immense pair of wings.' The creature appeared exhausted and could only fly short distances. The ranchers, armed with Winchester rifles, chased the creature for several miles, finally getting close enough to fire upon it and wound it. The creature turned on the men but as it was exhausted, they were able to keep out of its way. They shot the monster again, mortally wounding it. Upon examination they found that it was ninety-two feet long with a diameter of up to fifty inches. It had two feet, located a short distance in front of where the wings were joined to the body. The beak was about eight feet long, with strong, sharp teeth.

"The eyes were as large as dinner plates and protruded from the head. The wings measured seventy-eight feet, making the total length from tip to tip about one hundred and sixty feet. The wings were composed of a thick, nearly translucent membrane. The wings and body were hairless and featherless. The men cut off a small portion of the tip of one wing and took it home with them. One man then went into Tombstone to make preparations to skin the creature.

"The article claimed that the hide would be sent to eminent scientists for examination. The ranchers returned to the site accompanied by several 'prominent men' who would endeavor to bring the strange creature to town. To the best of my knowledge, this is the last that anyone ever heard of this creature. There is no men-

By one account, the mysterious Thunderbird's wingspan was said to be 160 feet, much larger than that of Quetzalcoatlus, the largest known pterosaur, which had a wingspan of 36–39 feet.

TIME TRAVEL: THE SCIENCE AND SCIENCE FICTION

tion of any photograph being taken. The article claims that the creature's wingspan was 160 feet. By comparison, Quetzalcoatlus's, the largest known pterosaur, had a wingspan of 36–39 feet, less than one-fifth that of the Tombstone flying monster.

"In later sections of this article we will be discussing the content of the Tombstone article. For now, it will suffice that we have established that there was a Thunderbird article in the *Epitaph* and that there was no photo accompanying it or mentioned in it."

Now let's try to figure out how the photo remains so elusive when crowds of people are absolutely sure they have seen it. The answer is a fascinating one and can be found in an interview I did with paranormal investigator and author Joshua P. Warren. He has a fascinating take on the photo and why no one has been able to find it. It does, of course, focus around time travel. Warren's words follow:

"I have thought, for a long time, about what I call the para-temporal loop hypothesis. At first glance it may not seem all that original, as it deals with the complexities that derive from potential time travel. The hypothesis is based upon one testable element. And that is, if ever, in all of the infinite future, any advanced species discovers how to travel back in time, will they do it?

"Let's say, hypothetically, that one million years from now—long after humans are gone, perhaps—there is a creature that dominates this planet that has evolved from the oceans. We'll call him Fish-Man. And Fish-Man is a great scientist and has discovered how to travel back in time. And so he does this. And let's say he goes back to the year 1920. Of course, he has to do his best to disguise his appearance, or else everyone will know what he is. And while he's back in 1920, he might be thinking how he could help, or even hurt, his own future existence."

> *"By traveling back in time, he has caused a para-temporal loop. It's a separate timeline that he continues to exist on."*

Josh continues: "And in doing so, he wants to be careful that he doesn't harm himself in the future. On the contrary, he might even try and enhance his future life by changing something in the past that will benefit him down the road. Or even something that will harm his enemies. Now, of course, Fish-Man will never know for certain if it's going to turn out right. If he screws up, then maybe he starts to vanish like the kid in *Back to the Future*. But if he does a good job, then he returns to his future and he finds that he has a better life. But by traveling back in time, he has caused a para-temporal loop. It's a separate timeline that he continues to exist on.

TIME TRAVEL: THE SCIENCE AND SCIENCE FICTION

"So, continuing with this thought experiment, let's say he gets back to his future, and things are better and brighter for him. And he doesn't want to jeopardize that by stating what he has done. But, he wakes up one day and everything is back the way it was before he tweaked it by going into the past. And he can't figure out what's happened."

Josh speculates that what might have happened is this: a billion years or so after Fish-Man is no more, along comes Bear-Man, who has also achieved time travel. Bear-Man heads back to an earlier time, such as 1915, and changes the timeline that Fish-Man had created. On this matter, Josh says: "Fish-Man has to then go back to 1910 to correct Bear-Man's adjustments, and so on, and so on. So, now, we have what seems like the plot of some bad sci-fi movie, where we have all these figures from different futures that are going back into the past and trying to tweak things to their benefit."

Is there any evidence to support this scenario? Josh thinks there just might be. He brings up the matter of that legendary Thunderbird picture. As we have seen, numerous people in the world of paranormal activity swear they saw, mainly in the 1960s, an amazing black-and-white photo of a legendary, huge Thunderbird of Native American history and folklore. The photo is said to have originated in the 1800s. Searches for the photo have not resulted in the discovery of anything relevant. Magazines like *Argosy*, *True*, and *Saga* have been scrutinized to the ultimate degree. The picture remains missing. How is it that this photo, which has been seen by so many investigators of the supernatural and the cryptozoological, now cannot be found anywhere?

> *"There might be a shifting timeline that we are passing through on a day-by-day basis.... One day the Thunderbird photo is in a magazine, and then when the timeline is played with again, it's no longer in the magazine."*

Josh answers that question with these words: "I get the impression that there might be a shifting timeline that we are passing through on a day-by-day basis. One day UFOs might be real, and the next they're not. The next day Bigfoot is running around your backyard, and the next day he doesn't exist. One day the Thunderbird photo is in a magazine, and then when the timeline is played with again, it's no longer in the magazine. And it may be that, day by day, hour by hour, or even minute by minute, small changes to the timeline are being made by these entities, or beings—perhaps the aforementioned Men in Black—coming back and constantly playing around with the past and the future. So, things we remember in the past, like the Thunderbird photo, suddenly no longer exist in the present."

Time-traveling Men in Black moving among us, erasing a photo for who knows what reasons, and doing so in an outfit that will work well across a century or more?

TIME TRAVEL: THE SCIENCE AND SCIENCE FICTION

It doesn't get much stranger than that. But it doesn't mean the story has no merit to it. The number of people who have seen the photograph of the massive, legendary bird are now in their dozens. Possibly there are many more. Not one of them can find the picture, even though they absolutely know they saw it.

CHAPTER 17

DOPPELGANGERS AND MULTIPLE TIMELINES

Ever heard of doppelgängers? You're about to. We're talking about exact doubles of ourselves and how these tie in with the matter of time travel—specifically in relation to the many controversies surrounding timelines. The *Britannica* website says that a doppelgänger is, "in German folklore, a wraith or apparition of a living person, as distinguished from a ghost. The concept of the existence of a spirit double, an exact but usually invisible replica of every man, bird, or beast, is an ancient and widespread belief. To meet one's double is a sign that one's death is imminent. The doppelgänger became a popular symbol of horror literature, and the theme took on considerable complexity. In *The Double* (1846), by Fyodor Dostoyevsky, for example, a poor clerk, Golyadkin, driven to madness by poverty and unrequited love, beholds his own wraith, who succeeds in everything at which Golyadkin has failed. Finally the wraith succeeds in disposing of his original."

Writing for the BBC, Zaria Gorvett has addressed this issue, too. In an article dated July 13, 2016, Gorvett writes: "Folk wisdom has it that everyone has a doppelgänger; somewhere out there there's a perfect duplicate of you, with your mother's eyes, your father's nose and that annoying mole you've always meant to have removed. The notion has gripped the popular imagination for millennia—it was the subject of one of the oldest known works of literature—inspiring the work of poets and scaring queens to death.

TIME TRAVEL: THE SCIENCE AND SCIENCE FICTION

"But is there any truth in it? We live on a planet of over seven billion people, so surely someone else is bound to have been born with your face? It's a silly question with serious implications—and the answer is more complicated than you might think. In fact until recently no one had ever even tried to find out. Then last year Teghan Lucas set out to test the risk of mistaking an innocent double for a killer.

"Armed with a public collection of photographs of U.S. military personnel and the help of colleagues from the University of Adelaide, Teghan painstakingly analyzed the faces of nearly four thousand individuals, measuring the distances between key features such as the eyes and ears. Next she calculated the probability that two peoples' faces would match.

"What she found was good news for the criminal justice system, but likely to disappoint anyone pining for their long-lost double: the chances of sharing just eight dimensions with someone else are less than one in a trillion. Even with 7.4 billion people on the planet, that's only a one in 135 chance that there's a single pair of doppelgängers."

A writer for *Ancient Origins* provides the following: "The mythology of spirit doubles can be traced back thousands of years and was present in many cultures of

"Folk wisdom has it that everyone has a doppelgänger; somewhere out there there's a perfect duplicate of you."

the past, holding a prominent place in ancient legends, stories, artworks, and in books by various authors. Perhaps the most well-known reference to spirit doubles or 'alter egos' is the doppelgänger, a word still used today to refer to a person that is physically or behaviorally similar to another person.

"Doppelgänger is a German word meaning 'double goer' and refers to a wraith or apparition that casts no shadows and is a replica or double of a living person. They were generally considered omens of bad luck or even signs of impending death—a doppelgänger seen by a person's relative or friend was said to signify that illness or danger would befall that person, while seeing one's own doppelgänger was said to be an omen of death. Some accounts of doppelgängers, sometimes called the 'evil twin,' suggest that they might attempt to provide advice to the person they shadow, but that this advice can be misleading or malicious. They may also attempt to plant sinister ideas in their victim's mind or cause them confusion. For this reason, people were advised to avoid communicating with their own doppelgänger at all costs."

Consider these words, too, from Tom Little at *Atlas Obscura*, who traced the doppelgänger phenomenon from ancient Egypt through the Victorian era. Having mentioned sightings in relation to English writers Percy Bysshe Shelley and Mary Shelley, Little added: "Doppelgänger encounters continued in the U.S., with several cases following a pattern of three sightings preceding death. Soon after his election in 1860, Abraham Lincoln saw his reflection doubled in the mirror, with one face beside the other with a ghostly pallor. He tried to show his wife the apparition, which appeared two more times when she was not present. While Mary Todd was at first worried about this behavior, she took the vision as a sign that he would serve two terms, but would die before the end of the second.

President Abraham Lincoln reportedly saw his doppelgänger in a mirror soon after his election in 1860.

"Lincoln is far from the only American to meet their double in the 1800s. The antebellum South was home to numerous accounts of fateful sightings, each under similar circumstances. Linda Derry, site director at the Old Cahawba ghost town in Alabama, is a curator of folklore originating from that region. She has uncovered several cases with similar circumstances as Lincoln's sightings."

And how about this, from *Paranormal Guide*? It states: "Although the combining of the words to form the term is relatively recent, a little over two centuries old, the idea of a spiritual, ghostly or demonic double (we'll just use the term 'ghostly' from here) of living people have existed for millennia. These doubles may at times be seen by others as performing a person's actions before the real person makes them, or they may be a shadow, performing the same movements, but after they have happened. They may also be seen in one's reflection; however the reflection is facing away. Much of the time a doppelgänger is viewed as an omen for a tragedy, illness or death of the person who is copied. If someone sees their own ghostly double it generally bodes very badly for them, and a number of quite famous people have had the ghastly experience."

"A physicist determined that if it were possible to travel to the past, it would likely involve the creation of a pair of ghostly twins that ultimately annihilate each other."

Now let's get to the issue of time travel and doppelgängers. Janey Tracey, in a 2015 article for the website *Outer Places*, states: "Traveling to the past at the risk of destroying oneself is a common staple of science fiction, not to mention the central concern behind the Grandfather Paradox. In real life, we still have no idea if time travel is really possible, but a physicist determined in a recent study that if it were possible to travel to the past, *it would likely involve the creation of a pair of ghostly twins that ultimately annihilate each other*" (italics mine).

It's this issue of twins being created as a result of traveling through time that may have a significant bearing upon the matter of doppelgängers. In other words, maneuvering through time just might result in the manifestation of a new version of ourselves every time a new timeline is created. And if one of those time travelers just happens to intrude upon *our* timeline, then—hey, presto—there's another incarnation of you, of me, and of who knows how many other people?

In other words, maneuvering through time just might result in the manifestation of a new version of ourselves every time a new timeline is created.

One of those who suspected that the doppelgänger phenomenon may have been caused by time-traveling doubles was the late paranormal authority and author Brad Steiger. He had a very good reason to suspect that was the answer to the mystery. And why might that be? Because Brad had a fascinating story of his own to tell. He shared the story with me just a few years before he died. It goes as follows:

"In the following cases I suspect a human agency involved in a strange campaign that was conducted regarding Steiger imposters who spoke at various conferences around the United States. On occasions the imposters allegedly conducted

TIME TRAVEL: THE SCIENCE AND SCIENCE FICTION

CARL SAGAN

Somebody imitating the paranormal authority Brad Steiger—perhaps a doppelgänger?—reportedly won a debate against the renowned astrophysicist Carl Sagan—whom Steiger himself never met.

themselves very well, thus making the whole enterprise of Counterfeit Steigers a seemingly futile project. On other occasions, the imposter's assignment was quite obviously to taint my reputation.

"On an unfortunate number of occasions, I received letters complaining of my outrageous and insulting behavior while speaking at a conference. There were claims that I had openly berated my audience, calling them stupid for accepting the very premise of UFOs. A close friend happened to arrive on the scene after one pseudo-Steiger had departed and tried his best to assure the sponsors of the event that the rowdy, disrespectful speaker could not have been the real Brad. In his letter, my friend warned me that he had visited a number of lecture halls where the imposter had damned his audiences. 'Someone seems out to damage your reputation,' he advised.

"In a most bizarre twist, dozens of men and women have approached me at various lectures and seminars, congratulating me about the manner in which I bested Dr. Carl Sagan in debate. The event allegedly occurred after a lecture when I happened to bump into the great scientist in a restaurant. The eatery, according to the witnesses, was crowded with those who had attended the seminar, and they egged on a debate between myself and Dr. Sagan. I mopped up the floor with him, countering his every argument against the reality of UFOs.

"The truth is that I never met Dr. Sagan, therefore, neither had I ever debated him. But from coast to coast, there are those who claim to have witnessed my triumphal bout. Even more individuals claim to have been in the audience when I delivered a rousing message from the Space Brothers in Seattle. Regardless of how often I deny that I was not in Seattle at that time and have never channeled the Space Brothers, those who were at that event are puzzled why I would deny my eloquence."

Multiple Brad Steigers wildly careering their collective way throughout the timelines? Don't bet against it.

CHAPTER 18

DISASTER IN THE SKIES

Not long before midday on January 28, 1986, the National Aeronautics and Space Administration suffered a horrifying catastrophe: the destruction of the space shuttle *Challenger*. Worse still, all of the crew lost their lives in the fiery explosion that took out the shuttle. They were pilot Michael J. Smith; payload specialists Gregory Jarvis and Christa McAuliffe; mission specialists Ellison Onizuka, Judith Resnik, and Ronald McNair; and the commander of the flight, Dick Scobee. Although the official verdict was that the *Challenger* disaster occurred as a result of wholly down-to-earth reasons, a wealth of conspiracy theories surfaced in the wake of the affair, all of which were carefully examined by none other than the Federal Bureau of Investigation. Before we get to the matter of time travel, (yes, there is a connection, and it involves time-traveling paranormal beings, as weird as it sounds), let us first take a look at what we know for sure, based upon NASA's careful study of all the evidence available.

The flight of *Challenger* on mission 51-L commenced at 11:38 a.m. EST. Almost immediately, it was all over. Just 73 seconds into the flight, a deadly explosion of oxygen and hydrogen propellants blew up the shuttle's external tank. This, as NASA noted, "exposed the Orbiter to severe aerodynamic loads that caused complete structural breakup. All seven crew members perished. The two Solid Rocket Boosters flew out of the fireball and were destroyed by the Air Force range safety officer 110 seconds after launch."

But how did such a thing happen? Inquiring minds, including the government, the media, and the general public, wanted answers. NASA responded with a detailed

Just 73 seconds into the flight of the space shuttle *Challenger* on January 28, 1986, a deadly explosion of oxygen and hydrogen propellants blew up the shuttle's external tank. The entire crew perished.

study of the evidence. NASA's investigative team noted in its report to the presidential commission investigating the accident:

"At 6.6 seconds before launch, the *Challenger*'s liquid fueled main engines were ignited in sequence and run up to full thrust while the entire Shuttle structure was bolted to the launch pad. Thrust of the main engines bends the Shuttle assembly forward from the bolts anchoring it to the pad. When the Shuttle assembly springs back to the vertical, the solid rocket boosters' restraining bolts are explosively released. During this prerelease 'twang' motion, structural loads are stored in the assembled structure. These loads are released during the first few seconds of flight in a structural vibration mode at a frequency of about 3 cycles per second. The maximum structural loads on the aft field joints of the Solid Rocket Boosters occur during the 'twang,' exceeding even those of the maximum dynamic pressure period experienced later in flight."

Just after liftoff, at 0.678 seconds into the flight, said NASA, "photographic data show a strong puff of grey smoke was spurting from the vicinity of the aft field joint on the right Solid Rocket Booster. The two pad 39B cameras that would have recorded the precise location of the puff were inoperative. Computer graphic analysis of film from other cameras indicated the initial smoke came from the ... aft field joint of the right Solid Rocket Booster. This area of the solid booster faces the External Tank. The vaporized material streaming from the joint indicated there was not complete sealing action within the joint.

"Eight more distinctive puffs of increasingly blacker smoke were recorded between .836 and 2.500 seconds. The smoke appeared to puff upwards from the joint. While each smoke puff was being left behind by the upward flight of the Shuttle, the next fresh puff could be seen near the level of the joint. The multiple smoke puffs in this sequence occurred at about four times per second, approximating the frequency of the structural load dynamics and resultant joint flexing. Computer graphics applied to NASA photos from a variety of cameras in this sequence again placed the smoke puffs' origin in the 270- to 310-degree sector of the original smoke spurt.

"As the Shuttle increased its upward velocity, it flew past the emerging and expanding smoke puffs. The last smoke was seen above the field joint at 2.733 seconds. At 3.375 seconds the last smoke was visible below the Solid Rocket Boosters and became indiscernible as it mixed with rocket plumes and surrounding atmosphere.

"The black color and dense composition of the smoke puffs suggest that the grease, joint insulation and rubber O-rings in the joint seal were being burned and eroded by the hot propellant gases. Launch sequence films from previous missions were examined in detail to determine if there were any prior indications of smoke of the color and composition that appeared during the first few seconds of the 51-L mission. None were found. Other vapors in this area were determined to be melting frost from the bottom of the External Tank or steam from the rocket exhaust in the pad's sound suppression water trays."

The report continued: "Shuttle main engines were throttled up to 104 percent of their rated thrust level, the *Challenger* executed a programmed roll maneuver, and the engines were throttled back to 94 percent. At approximately 37 seconds, *Challenger* encountered the first of several high-altitude wind shear conditions, which lasted until about 64 seconds. The wind shear created forces on the vehicle with relatively large fluctuations. These were immediately sensed and countered by the guidance, navigation and control system."

It was all to no avail, however. "At 45 seconds into the flight," NASA noted, "three bright flashes appeared downstream of the *Challenger*'s right wing. Each flash lasted less than one-thirtieth of a second. Similar flashes had been seen on other flights. Another appearance of a separate bright spot was diagnosed by film analysis to be a reflection of main engine exhaust on the Orbital Maneuvering System pods located at the upper rear section of the Orbiter. The conclusion was that the flashes were unrelated to the later appearance of the flame plume from the right Solid Rocket Booster.

"Both the Shuttle main engines and the solid rockets operated at reduced thrust, approaching and passing through the area of maximum dynamic pressure of

720 pounds per square foot. Main engines had been throttled up to 104 percent thrust and the Solid Rocket Boosters were increasing their thrust when the first flickering flame appeared on the right Solid Rocket Booster in the area of the aft field joint. This first very small flame was detected on image enhanced film at 58.788 seconds into the flight. It appeared to originate at about 305 degrees around the booster circumference at or near the aft field joint."

It was at the 72-second mark, NASA demonstrated, that "a series of events occurred extremely rapidly that terminated the flight." The agency stated: "At about 72.20 seconds the lower strut linking the Solid Rocket Booster and the External Tank was severed or pulled away from the weakened hydrogen tank permitting the right Solid Rocket Booster to rotate around the upper attachment strut. This rotation is indicated by divergent yaw and pitch rates between the left and right Solid Rocket Boosters."

Things had now reached the point of no return: overwhelming death and disaster were all but inevitable. The report continued: "At 73.124 seconds, a circumferential white vapor pattern was observed blooming from the side of the External Tank bottom dome. This was the beginning of the structural failure of the hydrogen tank that culminated in the entire aft dome dropping away. This released massive amounts of liquid hydrogen from the tank and created a sudden forward thrust of about 2–3 million pounds, pushing the hydrogen tank upward into the intertank structure. At about the same time, the rotating right Solid Rocket Booster impacted the intertank structure and the lower part of the liquid oxygen tank. These structures failed at 73.137 seconds as evidenced by the white vapors appearing in the intertank region.

"Within milliseconds there was massive, almost explosive, burning of the hydrogen streaming from the failed tank bottom and the liquid oxygen breach in the area of the intertank. At this point in its trajectory, while traveling at a Mach number of 1.92 at an altitude of 46,000 feet, the *Challenger* was totally enveloped in the explosive burn. The *Challenger*'s reaction control system ruptured and a hypergolic burn of its propellants occurred as it exited the oxygen-hydrogen flames. The reddish brown colors of the hypergolic fuel burn are visible on the edge of the main fireball. The Orbiter, under severe aerodynamic loads, broke into several large sections which emerged from

NASA's official explanation of the *Challenger* disaster involved "a failure in the joint between the two lower segments of the right Solid Rocket Motor." However, the FBI took various reports from sources who claimed knowledge of sabotage.

the fireball. Separate sections that can be identified on film include the main engine/tail section with the engines still burning, one wing of the Orbiter, and the forward fuselage trailing a mass of umbilical lines pulled loose from the payload bay."

NASA's conclusion on the affair reads as follows:

"The consensus of the Commission and participating investigative agencies is that the loss of the Space Shuttle *Challenger* was caused by a failure in the joint between the two lower segments of the right Solid Rocket Motor. The specific failure was the destruction of the seals that are intended to prevent hot gases from leaking through the joint during the propellant burn of the rocket motor. The evidence assembled by the Commission indicates that no other element of the Space Shuttle system contributed to this failure. In arriving at this conclusion, the Commission reviewed in detail all available data, reports and records; directed and supervised numerous tests, analyses, and experiments by NASA, civilian contractors and various government agencies; and then developed specific failure scenarios and the range of most probable causative factors."

With regard to the crew, according to a second report authored by biomedical specialist Joseph P. Kerwin of the Johnson Space Center in Houston: "The findings are inconclusive. The impact of the crew compartment with the ocean surface was so violent that evidence of damage occurring in the seconds which followed the disintegration was masked. Our final conclusions are: the cause of death of the *Challenger* astronauts cannot be positively determined; the forces to which the crew were exposed during Orbiter breakup were probably not sufficient to cause death or serious injury; and the crew possibly, but not certainly, lost consciousness in the seconds following Orbiter breakup due to in-flight loss of crew module pressure."

Although NASA's official conclusion was that the destruction of *Challenger* and the deaths of the crew members were the collective result of a terrible accident, in no time at all, conspiracy theories surfaced to suggest the event was not the accident that NASA claimed it was. Some of these conspiracy theories reached the very heart of the Federal Bureau of Investigation. Interestingly, the FBI did not ignore or write off the claims. Instead, the Bureau launched concerted investigations to get to the truth. We know this because the FBI's lengthy file on the *Challenger* conspiracy has now been declassified, thanks to the provisions of the Freedom of Information Act.

Less than 24 hours after the *Challenger* explosion took place, the office of William H. Webster, then the director of the FBI, received a memorandum from the agency's office in Boston, Massachusetts. It was a memo that described something disturbing and controversial. Barely 48 hours before the shuttle was destroyed, a reporter at the city's Channel 7 news took a phone call from an anonymous man who claimed that, according to the FBI's files, "he was part of a group of three people who were going to sabotage the Shuttle, causing it blow up and kill all aboard."

Boston-based FBI agents wasted no time in hitting the offices of Channel 7. The staff was extensively interviewed as the Bureau sought to gather all the available facts. Unfortunately, they were scant. They revolved around the caller's claims that "horrible things" were about to befall NASA and the *Challenger* crew, and that no fewer than "five people are going to be killed." Killed by whom was the big mystery facing the FBI.

At least, the Bureau assumed it would be a big mystery, given that the caller was anonymous and seemingly long gone. But that was not quite the case. For one of the agents, this was all too familiar, as a particularly notable, and now-declassified, FBI report shows. In part, the document reports: "During briefing of SAC [special agent in charge], ASAC [assistant special agent in charge], and appropriate supervisory personnel relative to aforementioned and employment of agent personnel, it was recalled that in September of 1985, a walk in complainant, of questionable mentality, had intimated that he had been responsible for the delay of previous Shuttles, plane crashes and other catastrophic events."

Agents who worked on the case well remembered the odd man, who clearly displayed far more than a few psychological issues. As a result, it didn't take them long to find and arrest the man. He was quickly subjected to what was described as a "five-day mental evaluation." It was clear to the FBI that the man was not faking his deranged mindset. As a result, he was released without charge, providing he underwent therapy and took whatever drugs the responding doctors determined he needed to take to maintain at least a degree of stability.

It must be said, though, that even some of the FBI agents on the case expressed their suspicions that there might have been more to the matter than met the eye. Yes, admittedly, the man had made a number of prior predictions about a terrible disaster concerning the *Challenger* space shuttle. But this one was unlike any of the previous ones: not only did the man correctly predict the destruction of the shuttle, but he also predicted it just two days before the disaster actually happened.

With the Boston case apparently solved, the FBI was far from done with space shuttle–based conspiracy theories, however. At the same time that agents of the Boston office of the FBI were pursuing leads on the destruction of *Challenger*, something of a similar nature was going down in California. The story is told in a summary document prepared by FBI agents in April 1986 after the investigation was finally closed. The document in question is titled "Space Shuttle Challenger, Information Concerning Launch Explosion, Kennedy Space Center, Florida, January 28, 1986," and dated April 18. It reads as follows:

TIME TRAVEL: THE SCIENCE AND SCIENCE FICTION

"On January 31, 1986, the FBI Resident Agency in Santa Ana, California was advised by [identity deleted] that he believes the Challenger exploded due to its being struck by laser beams fired from either Cuba or an aircraft. [Source] stated that a review of film footage of the explosion revealed brown puffs of smoke coming from the Space Shuttle just prior to the explosion. He stated leaks from the fuel tanks would produce white smoke, not brown smoke. [Source] said that the brown smoke would be produced each time the craft took a 'hit' by the laser beam, and the explosion occurred when the laser beam penetrated the skin of the craft."

The FBI took careful steps to speak with leading figures in the field of laser-based weaponry, both in the U.S. military and the private sector. Interestingly, just about everyone told the FBI that the scenario was, disturbingly, theoretically possible but was considered unlikely. Precisely why the scenario was dismissed when there was a near-unanimous consensus that such a thing could really be achieved is curious. Unfortunately, certain portions of the documents that have been declassified on this matter are significantly redacted, thus making it practically impossible to secure the full story.

Moving on from Massachusetts and California, the story then takes us to Dallas, Texas. It was early March when the Bureau's office in Dallas began investigating the claims of a man who worked in the movie industry. He believed that footage he recorded and carefully analyzed showed "something" flying through the sky and hitting one of the two boosters responsible for launching the shuttle, and "subsequently causing the explosion." FBI agents were sufficiently concerned to secure the footage—which they did after a lengthy interview with the man, whose name is deleted from the available files.

He believed that footage he recorded showed "something" flying through the sky and hitting one of the two boosters responsible for launching the shuttle, and "subsequently causing the explosion."

The matter was ultimately dismissed, although it should be noted that the files reflect the man was perceived as nothing less than a good, concerned citizen and not someone displaying mental issues or working to a suspicious agenda. The most bizarre story of all was still to come, however.

Demonstrating that the FBI's study of the *Challenger* explosion was very much a nationwide one, the story now takes us to Washington, D.C. It's a strange saga, made even stranger by the fact that, even today, nearly 30 pages of material on the affair remain classified, purportedly for having a bearing on the safety of the nation. It revolves around the claims of a woman who maintained two things: first, that the destruction of the space shuttle was the work of Japanese terrorists, and second, that her information on the matter was channeled into her mind by highly advanced, nonhuman entities.

The FBI's files detail the controversy surrounding the woman in question from nearly the start. The Bureau recorded in its documents on the case that the woman "claims to be in contact with certain psychic forces that provide her with higher information on selected subjects. She refers to these forces as 'Source' and when providing information from Source she often speaks in the collective 'we.' [She] claimed that she had come to Washington, D.C. to provide information concerning the *Challenger* Space Shuttle explosion on 1/28/86."

She did precisely that and provided the information on February 24, 1986. There was, however, far more to the matter of the beings known as the "Source." Paranormal entities, they had the ability to maneuver between time frames as they saw fit. They supposedly had the ability to manipulate timelines, too, and in the process create spin-off realities beyond ours, which must surely have had the FBI wondering what on earth—or off it—was going on.

According to FBI records, a woman claimed that paranormal entities who could maneuver between time frames, including two workers at the Kennedy Space Center and one of the *Challenger* astronauts, wished to destabilize the U.S. space program and caused the *Challenger* explosion.

TIME TRAVEL: THE SCIENCE AND SCIENCE FICTION

The woman's claims—provided by the "Source"—were controversial: she maintained that the terrorist group in question included two workers at the Kennedy Space Center and one of the astronauts who died in the disaster: mission specialist Ellison Onizuka. As the FBI agents working on the case listened carefully—and, perhaps, a bit dubiously too—they were told that the group in question had a deep hatred of the United States and, by destroying the shuttle, wished to destabilize the U.S. space program and American morale. Whether the woman's story was true or not, it is a fact that the U.S. public was indeed shocked to the core, and the space shuttle program was put on hold for 32 months.

When the agents asked the woman how the sabotage was achieved, they got a detailed answer: "The explosion was effected by a device placed inside the external fuel tank of the Shuttle. An individual whose description seems to match that of an engineer or technician placed this charge. The charge was triggered by a second saboteur using a hand-held transmitter while standing in the crowds watching the Shuttle lift-off. The individual matches the description of a guard or security person. The astronaut saboteur chose to die in the explosion as a sort of ritual death or 'cleansing.'"

As with all of the previous cases that the FBI had looked into, this one led nowhere—at least, that is the assumption since no arrests were made. The odd affair came to a complete halt just weeks after it commenced.

The destruction of the *Challenger* space shuttle, on January 28, 1986, remains to this day one of the worst moments in NASA's history. Whether it was a moment provoked by nothing stranger than a terrible accident or something filled to the brim with conspiracy theories and sinister, ruthless characters very much depends on whom you ask.

CHAPTER 19

MOTHMAN AND FUTURE EVENTS

There can be few people reading this who have not at least heard of the legendary Mothman of Point Pleasant, West Virginia, who terrorized the town and the surrounding area between November 1966 and December 1967, and whose diabolical exploits were chronicled in the 2002 hit Hollywood movie starring Richard Gere, *The Mothman Prophecies*, named after the book of the same title written by Mothman authority John Keel. Perhaps not all have made the connection between the Mothman and time travel, however.

As the story goes, a devil-like, winged monster with glowing red eyes came quite literally out of nowhere and, some say, instigated high tragedy and death. But what was the Mothman of Point Pleasant? And how did the legend begin? To answer those questions, we have to go back to the dark night of November 12, 1966, when five gravediggers working in a cemetery in the nearby town of Clendenin were shocked to see what they described as a "brown human shape with wings" rise out of the thick, surrounding trees and soar off into the distance.

Three days later, the unearthly beast surfaced once again. It was at the witching hour, a highly appropriate time, when Roger and Linda Scarberry and Steve and Mary Mallette, two young married couples from Point Pleasant, were passing the time by cruising around town in the Scarberrys' car.

As they drove around the old factory, the four were puzzled to see in the shadows what looked like two red lights pointing in their direction. These were no

A devil-like, winged monster with glowing red eyes, the Mothman appeared in Point Pleasant, West Virginia, in the 1960s and, some say, instigated high tragedy and death.

normal lights, however. Rather, all four were shocked and horrified to realize that the "lights" were the glowing, self-illuminating red eyes of a huge animal that, as Roger Scarberry would later recall, was "shaped like a Mothman, but bigger, maybe six and a half or seven feet tall, with big wings folded against its back."

Not surprisingly, the party fled the area at high speed. Unfortunately for the Scarberrys and the Mallettes, however, the beast seemingly decided to follow them: as they sped off for the safety of Point Pleasant, the winged monster took to the skies and shadowed their vehicle's every movement until it reached the city limits.

The four raced to the sheriff's office and told their astounding story to Deputy Millard Halstead, who later stated: "I've known these kids all their lives. They'd never been in any trouble and they were really scared that night. I took them seriously."

And even though a search of the area by Halstead did not result in an answer to the mystery, the Mothman would soon return.

Early on the morning of November 25, yet another remarkable encounter with the mysterious beast took place, as John Keel noted: "Thomas Ury was driving along Route 62 just north of the TNT area when he noticed a tall, grey manlike figure standing in a field by the road. 'Suddenly it spread a pair of wings,' Ury said, 'and took off straight up, like a helicopter. It veered over my convertible and began going in circles three telephone poles high.'"

> *"It was tall with big red eyes that popped out of its face.*
> *My husband is six feet one and this bird looked about*
> *the same height."*

TIME TRAVEL: THE SCIENCE AND SCIENCE FICTION

Keel reported that Ury quickly hit the accelerator. Nevertheless, Ury added: "It kept flying right over my car even though I was doing about seventy-five." Over the next few days, more sightings surfaced, including that of Ruth Foster of nearby Charleston, who saw the winged monster late at night in her garden. Foster said: "It was tall with big red eyes that popped out of its face. My husband is six feet one and this bird looked about the same height or a little shorter, maybe."

Needless to say, the local media had a field day with the story. Tales of what were referred to as the "Bird-Monster" hit the headlines, and both the skeptics and the police ensured that their views and opinions on the matter were widely known.

Dr. Robert L. Smith, associate professor of wildlife biology in the West Virginia University's Division of Forestry, expressed his firm opinion that Mothman was nothing stranger than a large sandhill crane. This hardly satisfied the witnesses, however. In response to Dr. Smith's assertion, Thomas Ury said: "I've seen big birds, but I've never seen anything like this."

As for the local police, they offered stern warnings to any and all would-be monster hunters contemplating seeking out the mysterious creature, as the *Herald Dispatch* newspaper noted: "Sheriff [George] Johnson said he would arrest anybody caught with a loaded gun in the area after dark [and] warned that the scores of persons searching the abandoned powerhouse in the TNT area after dark risked possible serious injury."

In the weeks and months that followed, further encounters with the bizarre beast were reported; however, they were overshadowed by a tragic event that occurred on December 15, 1967. It was on that day that Point Pleasant's Silver Bridge (so named after its aluminum paint), which spanned the Ohio River and connected Point Pleasant to Gallipolis, Ohio, collapsed into the river, tragically claiming 46 lives. Interestingly, after the disaster at the Silver Bridge, encounters with the Mothman largely came to a halt.

And while a down-to-earth explanation most certainly circulated—namely, that a fatal flaw in a single eyebar in a suspension chain was the chief culprit—many linked the cause directly to the ominous and brooding presence of the accursed Mothman.

As for the angle of time travel and Mothman, it hinges on a connection to certain prophecies. John Keel did not title his book *The Mothman Prophecies* for no reason. Keel himself foresaw a terrible disaster enveloping the city, resulting in terrible carnage and death in the dozens. Admittedly, Keel wasn't entirely sure what was looming on the horizon, only that something was.

In 2002 a big-bucks movie version of *The Mothman Prophecies* was made, starring Richard Gere. Although the movie was fictionalized, it was fairly close to the

TIME TRAVEL: THE SCIENCE AND SCIENCE FICTION

Many linked the cause of the Silver Bridge collapse in December 1967 directly to the ominous and brooding presence of the accursed Mothman. Others suggested the sightings were actually warnings of future disaster.

reality of the sinister situation and included a reference to certain prophecies, hence the title of both the book and the movie. The *Chasing the Frog* website asks if there really were prophecies around town in the 1966–67 era. There certainly were, as the site shows. About a local girl named Connie Mills, it asks the question: "Did Connie Mills really have a dream about drowning while surrounded by Christmas gifts?"

> *In short, people in and around Point Pleasant were seeing future events.*

The answer: "Not exactly. A dream prophecy was reported and the event happened, however, it was not the same premonition as in the movie. Mary Hyre, a newspaper reporter that often accompanied Mr. Keel in Point Pleasant investigations, dreamt that there were a lot of people drowning in the river and Christmas packages were floating everywhere in the water (*The Mothman Prophecies* book). Her counterpart in the movie, Connie Mills (Laura Linney), describes a dream in which she herself is drowning in an ocean, surrounded by floating Christmas presents."

What this demonstrates is that before the terrible tragedy at Point Pleasant's Silver Bridge occurred, people were having glimpses, in the format of dreams and

TIME TRAVEL: THE SCIENCE AND SCIENCE FICTION

prophecies, of events and deaths that had not yet occurred but soon would. In short, people in and around Point Pleasant were seeing future events.

The story is not yet over. There is another part that ties in with time travel, as we examine in the following chapter.

TIME TRAVEL: THE SCIENCE AND SCIENCE FICTION

CHAPTER 20

A TIME TRAVELER
COMES IN FROM THE COLD

The next part of our story of strange happenings in the Ohio River valley revolves around a man named Woodrow Derenberger and a decidedly sinister character who went by the moniker of Indrid Cold. Many saw him as an alien. One famous writer, though, thought he was a time traveler. The story of Derenberger and Cold was provided to me by the late Susan Sheppard, a local who knew the story inside and out. Susan very generously prepared the following statement for me, which I now offer to you. It begins on November 2, 1966, and, as you'll find, there is a connection with our old friends, those curious time travelers. Over to Susan:

Everyone called him "Woody." It was shortly after six o'clock in the evening when Woody Derenberger was driving home from his job as a sewing machine salesman at JCPenney in Marietta, Ohio, to his farmhouse in Mineral Wells, West Virginia. The ride home was overcast and dreary. It was misting a light rain.

To his astonishment, what Derenberger thought was a truck was a charcoal-colored UFO without any lights on.

As Derenberger came up on the intersection of I-77 and Route 47, he thought that a tractor trailer truck was tailgating him without its lights on, which was unnerving, so he swerved to the side of the road. Much to his surprise, the truck appeared to take flight and seemed to roll across his panel truck. To his astonishment, what Derenberger thought was a truck was a charcoal-colored UFO without any lights on. It touched down and then hovered about 10 inches above the berm of the road. Much to Derenberger's surprise, a hatch opened and a man stepped out, looking like "any ordinary man you would see on the street—there was nothing unusual about his appearance."

Except the man was dressed in dark clothing and had a "beaming smile." As the man proceeded to walk toward Derenberger's panel truck, the "craft" jetted up to about 40 feet in the air, where it floated above the highway. What happened next was unsettling, because as the darkly dressed man came up toward the vehicle, Woody Derenberger heard the words, "Do not be afraid. I mean you no harm. I only want to ask you a few questions." Derenberger did become afraid, because as the man spoke to Woodrow, his lips did not move. The man then moved to the opposite of the truck and told Derenberger to roll down his window so they could talk better, which he did. Next, what formed in Derenberger's mind were the words, "Now you can speak, or you can think. It makes no difference. I can understand you either way." This is what the dark man said.

Later, when Derenberger was questioned on local live television, he was scrutinized over what seemed a contradiction because if the dark man communicated

Woodrow Derenberger described Indrid Cold as about 35 years old, trim, about six feet tall, with dark eyes and dark hair slicked straight back, and wearing a long, dark coat.

TIME TRAVEL: THE SCIENCE AND SCIENCE FICTION

through a type of mental telepathy, why would Derenberger need to roll down his window to talk? Wouldn't it be easier just to talk mentally?

Woodrow Derenberger explained it was because Indrid Cold wanted to look directly at him as they spoke, and he felt that, really, Cold wasn't so interested in what was said but more interested in keeping up a communication with him. To Derenberger, that seemed the entire point. Derenberger also noted that when Cold stared into his eyes, it was as if Cold knew everything about him, and also, if he could only let go of his fear and do the same, Derenberger felt he would also know and understand all about Cold. In any event, Cold spoke through the passenger-side window the entire time.

The physical description of Cold was commonplace. Derenberger described him as about 35 years of age, having a trim build, about six feet tall, 185 pounds, with dark eyes and dark hair slicked straight back. Cold wore a long, dark coat, and Derenberger was able to glimpse the fabric of his "uniform" that glistened beneath the coat. He also described Cold as having a "tanned complexion." Throughout the conversation Cold kept a frozen smile and curiously hid his hands beneath his armpits most of the time.

Cold did, however, point at the city lights above the distant hills of Parkersburg, West Virginia, and asked Mr. Derenberger, "What do you call that over there?" Derenberger said, "Why, that's Parkersburg and we call that a city." Cold responded, "Where I come from we call it a gathering." Cold later added the curious statement that "I come from a place less powerful than yours." As the men talked, cars passed under the craft, which drifted above the road. The drivers and occupants were seemingly unaware of the spaceship being there. After all, there were no lights that could be seen. Cold then asked about Parkersburg, "Do people live there or do they work there?"

Derenberger answered, "Why, yes, people live and work there." Cold interjected, "Do you work, Mr. Derenberger?" (Woodrow had told Cold his name as the conversation began.)

Derenberger answered, "I am a salesman. That's what I do. Do you have a job?" Cold answered, "Yes. I am a searcher."

After that the conversation became mundane. Cold seemed to notice Woodrow Derenberger was scared and commented on it. Mr. Derenberger claimed that Cold asked him, "Why are you so frightened? Do not be afraid. We mean you no harm. You will see that we eat and bleed the same as you do," and then added an emotive note, "We only wish you happiness," which Cold said to the frightened man more than once.

When Mr. Derenberger was interviewed later on live television on WTAP-TV, he attributed this puzzling statement to Indrid Cold: "At the proper time, the authorities will be notified about our meeting and this will be confirmed." The entire conversation took between five and ten minutes. Then Indrid Cold looked inside

Woody's car with his ever-present smile and said, "Mr. Derenberger, I thank you for talking to me. We will see you again."

As soon as he said that, the spaceship immediately came back down and floated about 10 inches off the road. A hatch opened, and a human arm extended to pull Cold up into the craft. The ship then jetted up into the air about 75 feet, made a fluttering noise, and then shot away at a very high rate of speed. For a few moments, Woodrow Derenberger sat stunned. Finally, he started up his car and drove to his farmhouse in Mineral Wells, where his wife met him at the door. By now it was shortly before seven o'clock.

Mrs. Derenberger later said that Woodrow "could not have been any whiter if he had been lying in a coffin." The stories vary, but from Mr. Derenberger's account, his wife is the one who called the West Virginia State Police, or at least she dialed the phone. Woodrow Derenberger gave them a brief report of what he claimed had happened. It is interesting to note that in the initial report, Derenberger called the alien "Cold," but he did not mention "Indrid" until later.

The next day Derenberger attempted to go back to work but was sidetracked when he agreed to a live television interview about his experience on the previous night with a UFO with WTAP-TV, the NBC affiliate in Parkersburg, housed in a small building not much bigger than a garage. The interview took place shortly before noon, and Woodrow Derenberger was grilled by veteran reporter Glenn Wilson, city police chief Ed Plum, and other local law enforcement, including the head of the Wood County Airport. Representatives from Wright-Patterson were in route to interview Derenberger, but whether that came about is not known.

The interview went on for about two and a half hours. The live part of the broadcast was under an hour long, and then the television cameras were turned off. The interview continued off the air for another hour or so. During that time, Derenberger drew a picture of the spacecraft, which he described in his thick West Virginia accent as charcoal grey, with no lights, and looking like an "old-fashioned chimney lamp."

Woody Derenberger described the spacecraft in which Indrid Cold flew away as charcoal grey, with no lights, and looking like an "old-fashioned chimney lamp."

Probably one of the most curious statements Woodrow Derenberger made about his meeting with Cold was this: "And then Cold said to me, we will see you again …" then his voice trails off. Police Chief Ed Plum asked, "Do you really believe you will see him again?" Derenberger then answered, "I think so … I believe I will … I don't know … because that's what I am afraid of."

After that interview, Derenberger's life transformed drastically, and not for the better. He changed jobs, developed marital problems, and clung to his church for a while. Then came the strange visits from men dressed in black clothing whom Derenberger suspected to be some kind of hidden government group of spies or maybe even the Mafia. He wasn't sure; they just spooked him. They would arrive at his house, ask Derenberger simple questions (some having to do with his UFO experience), and then the Men in Black would act in a threatening manner.

But nothing was as incredible as the return of Indrid Cold. At least, this is what Woodrow Derenberger claimed. He said that Cold visited him many times at his farmhouse in Mineral Wells. At one point, Derenberger came up missing for almost six months and later said he had been "with the aliens." The local population finally became skeptical. The sewing machine salesman's tale grew more and more far-fetched. Derenberger even claimed to have been impregnated by the aliens. In 1967, Derenberger claimed to have visited Indrid Cold's home planet of Lanulos, where its residents walked around wearing no clothing. He said the aliens lived in a galaxy called Ganymede, where everything was peaceful and there was no war. People began to snicker.

Still, there were odd flashing lights in the sky almost nightly, and the curiosity seekers stalked not only Derenberger's modest farmhouse but an area called Bogle Ridge, not far from Mineral Wells, where the aliens were claimed to land. The ridicule became too much. Derenberger, with his family, moved from the area and stayed away for decades. He returned to Wood County in the 1980s and died in 1990. Woodrow Derenberger was finally laid to rest at Mount Zion Cemetery in Mineral Wells, West Virginia.

John Keel was not a believer in Woodrow Derenberger's UFO story, so it's mysterious why he would have made it such a big part of *The Mothman Prophecies* book. In *The Mothman Prophecies* movie, the character Gordon Smallwood is based on Woodrow Derenberger, but the Wood County man most often appeared in a suit and not overalls. A few elements to his story make it believable that, initially, something of an extraordinary nature happened to him. First of all, his account predates the Mothman sightings by 12 days. Derenberger would have had to have been a prophet to know what was about to happen next, making his story even more extraordinary. His family explains that they believe something of an otherworldly nature initially happened, but he added to the tale to sell books when he self-published a book called *Visitors from Lanulos* in 1971.

There are a few other accounts that add some believability to key aspects of Woodrow Derenberger's fantastic story. An elderly man driving south of Parkersburg

TIME TRAVEL: THE SCIENCE AND SCIENCE FICTION

on I-77 reported seeing a man by the side of the road who met Indrid Cold's description and who tried to flag him down. The gentleman slowed down, but when the darkly clad man headed for the passenger door of his car, the senior citizen became frightened and drove away. "There was something off about that character," the old man later told Glenn Wilson of WTAP-TV.

> **In the same section of the newspaper was a smaller**
> **article about a complete power outage that happened**
> **in South Parkersburg at precisely the same time**
> **Derenberger claimed he was interrupted by Indrid**
> **Cold's spaceship.**

I also ran into something curious when I was researching stories for my ghost tour back in 1996. I found the news article about Woodrow Derenberger's UFO tale in the *Parkersburg News & Sentinel* dated November 4, 1966, where the story was on a front section of the newspaper. The account read "Local Man Stopped by UFO." In the same section of the newspaper, right beside it, was another smaller article about a complete power outage that happened in South Parkersburg at precisely the same time Derenberger claimed he was interrupted by Indrid Cold's spaceship. South Parkersburg borders the community of Mineral Wells. An energy disturbance in Mineral Wells would likely also affect South Parkersburg.

I did monthly live horoscopes for an astrology segment on WTAP-TV in Parkersburg for a number of years and knew Glenn Wilson, the man who had interviewed Woodrow Derenberger in November of 1966. When Glenn Wilson retired around 2001, he had something he thought I might like to have: they were the original reel-to-reel tapes of his live interview with Derenberger on November 3, 1966. These were audio tapes of more than two hours long, and on them was written "UFO TAPES November 1966." Wilson almost threw them away, he told me, because he felt Derenberger had given Parkersburg a "black eye," making the city a laughingstock over his alien visits and extraterrestrial pregnancy claims. Wilson said there was also a video of the live interview and a drawing Derenberger had produced of Indrid Cold's spaceship, but both went missing. Wilson assumed the cleaning lady had thrown them out with the garbage.

The infamous UFO interview had not been listened to for about 35 years, and hearing the tapes for the first time was quite remarkable. Musician and Emmy winner David Traugh, who owned a recording studio in Parkersburg, transferred the rare interview to a cassette tape in the summer of 2001. Woodrow Derenberger had come back to life to tell his story all over again. One could hear him rap his knuckles on the table during the interview for emphasis and listen as he faltered a bit. Yet he was consistent in everything he said.

Later, I burned the interviews onto a CD and presented them to author John Keel in 2003 at the only Mothman Festival (held each year in Point Pleasant) he attended. He commented that he didn't even know the interviews existed and seemed skeptical that they were real. I assured him they were genuine and said I hoped he would enjoy them. The last thing John Keel said to me was, "I hope you make some money off of these." (I haven't, really; I've given away more CDs than I've sold. In any event, that wasn't the purpose anyway.)

However, before all of this, back in 2001, I was doing a book signing for my astrology book that was published by Kensington Publishing when a young man in a long, dark coat appeared at the bookstore. He was dressed in dark clothing, and he was about six feet tall with dark hair combed straight back. He had a medium build and dark eyes. The man appeared to be in his mid-30s and introduced himself as "Billi" when he had me sign a book.

Billi was very good-looking. In fact, he resembled the actor Richard Gere, and I thought he had a Slavic or part Native American appearance. But what struck me instantly was that Billi didn't seem very intelligent. In fact, he acted rather dumb.

Billi picked up my book and asked me, "Where did you find this book?"

I answered, "Well, I wrote it, and this is my book signing."

Billi turned the book over. "Does it have to do with stars?"

I said, "Yes, your sign, like when you were born. The month and date tells what your sign is."

He looked puzzled and commented, "I was born in November of 1966."

I then said, "That would make you probably a Scorpio, intense in nature, interested in subjects others may feel a bit off-putting."

Billi looked thoughtful. "I assist in brain surgeries. I have witnessed many. I like the way the brain looks when the skull is open."

"Oh, you're a doctor?" I asked. (My inner voice is now saying, *How can this man be a doctor?*)

Billi answered, "No. I am a helper."

"You mean a nurse?" I asked.

"Yes, I am a nurse, I guess." Billi giggled. (Inner voice: *There is no way this guy made it through nursing school.* Also, brain surgeries were not performed in the two local hospitals. Brain surgery patients were usually sent to Morgantown, Pittsburgh, or Columbus.)

"Oh," I responded. (Inner voice asks, *Why is this man lying to me?*)

TIME TRAVEL: THE SCIENCE AND SCIENCE FICTION

"I think I am going to buy this. Will you write your name on it for me?" Billi clasped the book to his chest and then took it to the cash register. He pulled some bills out of his pocket, paid for the book, and brought it back to me.

"Put your name there," Billi commanded as he sat the book down in front of me.

I said, "Okay ... but yours, too. What's your name?"

Billi said, "My name is Billi."

"As in B-I-L-L-Y?" I spelled the name out.

"No," said Billi, "there is an 'I' at the end of my name."

I signed the book "To Billi" and asked, "Is that the right way?"

Billi said, "Yes. That's correct."

"Well, there you go, and thanks so much." I handed Billi his book.

"Thank you for talking. I have to go now ... to go find my brother ... he went somewhere. He may be lost, I think." Billi glanced over his shoulder, picked up the astrology book, and vanished through the doorway of the bookstore.

When writing on speculative subjects, including astrology, you meet people from all walks of life, and an odd person appearing can be normal in the strange,

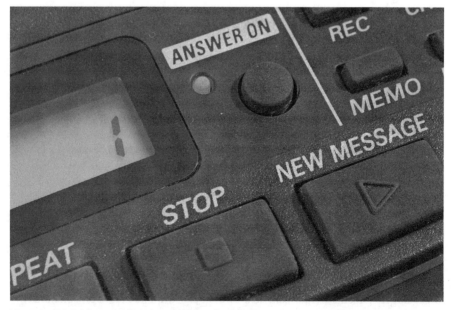

Was it Indrid Cold, now calling himself Billi, who left a bizarre message on Susan Sheppard's answering machine in August 2001?

speculative world, so I didn't give Billi much thought other than to remember the peculiar encounter.

It was now August of 2001, so it was time to begin preparation for my seasonal ghost tour, and that meant putting a new message on my answering machine. One night I walked in to find my answering machine's red light blinking (those were the days) and played back the message, which turned out to be a hissing voice saying the word "Hi!" But the word was drawn out, and the raspy greeting sounded more along the lines of: "Hiiiiiiiiiii-eeeeeeeee-yaaaaaaaah!" It was really a long hiss.

I'm used to prank calls, so I deleted the recorded message and carried on. There was something about the voice that didn't seem exactly human; however, some people can do bizarre acrobatics with their voices, and as with my meeting with Billi, my feeling was mainly a "welcome to my strange world" sentiment that I brush off and partly forget.

In the following weeks a catastrophe would hit the United States on September 11, 2001, when the World Trade Center in New York City, the Pentagon, and Shanksville, Pennsylvania, were attacked and 2,996 innocent individuals lost their lives. Like most Americans, I tried to move past my own shock and fulfill my obligations, and that would be preparing for another ghost tour coming up in little over a month. I thought maybe one new twist on the ghost tour might be playing some of the Derenberger/Indrid Cold interviews for the crowd. It would be not only an escape for everyone but also entertaining. In late 2001, aliens, as well as ghosts, were now pretty low on the list of scares.

The 2001 fall season grew cold rather quickly, and tourgoers were soon wearing winter coats in October. One night, I looked over as I told the Indrid Cold story and was about to play the Derenberger tapes to the crowd when I glimpsed a man dressed in a long, dark coat who stood apart from the rest of the crowd.

It was Billi.

Absorbed in my stories, Billi remained polite throughout the evening and trailed behind at the end of the crowd. Back at the hotel as the tour ended, Billi said, "I like the way you tell stories." Then, like a human-sized crow, he disappeared through the hotel's front doors. I watched through the window as Billi paused at the wait-and-walk sign outside and then crossed the street in the chilled fall air. Billi attended the ghost tour about two or three times that season. He was always quiet, and he was always alone.

In November of 2001, a phone call woke me up about 2 a.m. I recognized the voice. It was the same raspy voice that left a message on my answering machine earlier, with the exact same message: "Hiiiiiiiiiii-eeeeeeeee-yaaaaaaaah!" (As in "Hi!") The inhuman voice unnerved me, so I slammed the receiver down but instantly regretted it. Perhaps I could find out who was pranking me.

TIME TRAVEL: THE SCIENCE AND SCIENCE FICTION

Curiosity got the best of me, so I dialed *69, which in this area of the country is a way to learn the last telephone number that called. I don't remember the exact number, but I immediately found out the number was from a Point Pleasant, West Virginia, line. I only know two residents of Point Pleasant well enough to call me on the telephone: Jeff Wamsley of the Mothman Museum (and Festival) and my brother-in-law's sister. The number belonged to neither of them. I decided I would risk being rude, so I dialed the number even though it was past 2 a.m. All I had to do was punch the number "1" on my phone as a call back, and the phone would ring. I must admit, I was not surprised when an electronic voice clicked on and said, "The number you have called has been disconnected and is no longer in service." Yet the number had called me five minutes earlier.

I sat down in my chair, shaken. There were perceptions in the back of my mind I knew I had been repressing because I considered them impossible. I thought of the appearances of Billi and how he said he had been "born in November of 1966." And how Billi fit perfectly the description of Indrid Cold, with his long, dark coat, his tanned complexion, and hair combed straight back. And if they were the same, Cold had not aged and still remained 35 years old. I also remember how diligently Billi had listened to the Woodrow Derenberger tapes, as if he had a secret, but never commented. However, this might all have been just an overwhelming coincidence, so I left it at that.

On Thanksgiving weekend of each year, the local Smoot Theater features a production of *A Christmas Carol* that is put on by the Missoula Theater of Nebraska. It's a grand show that local people look forward to attending year after year. My family at the time was no exception. We looked forward to attending the Friday night presentation in 2001 following Thanksgiving, which, as always, was a Thursday. My family had good seats, and we waited as other members of the audience filled the theater. To my right was my then husband, and to my left was an empty chair that was an aisle seat. While my eyes scanned the darkened theater, I heard a slick rustling and noticed that someone sat down beside me. I turned.

It was Billi, still wearing his long black coat. I'd never seen Billi change into anything else.

As I sat there, I wanted to fill my husband in on what was going on, but I was simply too frightened to. Billi turned to me and said in his naive way,

The mysterious Billi appeared at a local production of *A Christmas Carol*—as it happens, another tale that involves time travel.

"I didn't think I was ever going to get here." I nodded but said nothing. Soon dancers and singers in costume filled the stage. Billi leaned in and whispered, "What do you call what they are wearing up there?"

I finally answered, "Those are their costumes."

Billi commented with a naive amazement, "Aren't those colors beautiful?" I nodded.

The dark-dressed man, who now seemed almost a friend, squirmed in his seat, and before the play was over Billi left. After the play had concluded, my 10-year-old daughter commented that she noticed Billi's black coat was made out of an unusual material, like the waterproof fabric on a tent.

Holiday shopping and the wrapping of presents means lots of pizza deliveries to my home in December. One evening of gift-wrapping brought a surprise to the door. We had ordered a pizza for dinner, having decided to decorate the tree and then maybe watch Christmas movies.

When the pizza man knocked on the door, I opened it to find Billi standing there holding a pizza box. I handed him a 20-dollar bill, told him to keep the change, and without acknowledging that I knew him, I quickly slammed the door and locked it. By this time I had lost my appetite, so I didn't eat the pizza and said nothing about it to my family. My family survived the pizza, and I survived another surprise visit from Billi.

Around this same time period, my best friend called me and said three of her grandsons had Chiari syndrome, and two of them would need brain surgery. The family would have to travel out of the area for the boys to get it. I could not help but remember Billi's far-fetched comments about assisting brain surgeries only a few months earlier.

A local art association called Artsbridge used to put on what was called its Chair Auction, where area artists painted used furniture that was auctioned off to help the organization with funding in art education. I was one of the local artists invited to paint a piece of furniture, which I did. I painted an old telephone stand with a Halloween theme and even added a ceramic vintage pumpkin that looked like a dreaming, sleeping child.

This would have been late summer of 2002. On the night of the auction, we assembled at the local art center to enjoy the refreshments and watch our art pieces being auctioned off. I was sitting with my friend whose grandsons had just had brain surgery for their Chiari syndrome. It was still not long after 9/11, and everyone needed a lift.

As the second-floor banquet room of the art center began to fill with people, I noticed Billi in his dark coat enter the room. But he wasn't alone. Billi arrived with a surreal-looking blond companion over six feet tall with pointy, cone-like breasts.

Wearing a shade of bright coral lipstick, the woman resembled a late '50s to early '60s pinup model. She was dressed mostly in black. I poked my friend and commented, "That's Billi!" I pointed in their direction. My friend's eyes widened, and you would have thought she was looking at the chain-rattling ghost of Ebenezer Scrooge. But it was Billi, all right, and he wasn't a ghost.

The auction began. Soon, my art piece came up for bids. The bidding grew fast and heated. My friend's eyes darted back and forth, and then her gaze fixed on me. "He's bidding on you," she whispered. (I dared not look.)

Not surprisingly, Billi was the highest bidder and bought my Halloween-themed, vintage telephone stand. I was stunned for the third or fourth time. When was this going to end? My friend had the composure to go downstairs where items were being paid for, so she followed Billi and his blond bombshell to the checkout area. My friend came back upstairs and said to me, "Susan, Billi paid for your art in cash. He then clasped the stand to his chest and ran out of the art center as if he scored some great prize."

After this, I never saw Billi again. It makes me wonder if my art is now being looked upon with quizzical stares by humanoids dressed in black, far, far away in a galaxy called Ganymede, on a planet named Lanulos, perhaps a place that is "less powerful that yours," one inhabited by a guy called "Billi" or maybe even "Indrid Cold."

Thus ends Susan Sheppard's tale. Now we get to the matter of Indrid Cold and time travel.

Robert Davis was someone who, before his passing in 2019, spent a lot of time digging deep into the early years of the UFO subject, the 1940s to the 1970s. Davis was noted for his huge amount of correspondence with fellow UFO seekers. One of them was a man named Gray Barker. He wrote the very first book on the Men in Black, 1956's *They Knew Too Much about Flying Saucers*. In correspondence in 1981, Barker confided in Davis that—after looking closely into the story of Indrid Cold, and after having written his own book about Mothman and Indrid Cold (*The Silver Bridge*, published in 1970)—he suspected that Cold was not an alien but a time traveler. Maybe even a time-surfing historian taking a peek around late twentieth-century America. Frustratingly, Barker didn't expand on how or why he thought that Cold was from the future, only that Barker had received a "message" from Cold himself. Such a thing is not impossible, as the matter of Indrid Cold and time traveling appears in another situation, as we shall see right now.

He suspected that Cold was not an alien but a time traveler. Maybe even a time-surfing historian taking a peek around late twentieth-century America.

TIME TRAVEL: THE SCIENCE AND SCIENCE FICTION

John St. Clair is a graduate of Cornell University in mechanical engineering. There he became very interested in Einstein's special theory of relativity. Before this, he won first prize in the Puerto Rico National Science Fair with a project in solar energy. He nearly burned up in flames the Fresnel lens furnace when removing it from the van below the noonday sun. He built the first human-occupied ground-effect hovercraft in Latin America. Working at Kodak's Optical and Apparatus Division, John obtained several inventions in film and slide projectors. As it happens, John has something fascinating to say about time travel, Indrid Cold, and timelines. St. Clair says:

I went to see the Will Smith and Tommy Lee Jones movie about the Men in Black (MIB). The movie was entertaining, yet the portrayal of the MIB as secret government agents I thought was incorrect. Then I read Nick Redfern's book *The Real Men in Black* that documented many cases about witnesses who had come in contact with these mysterious MIB beings. Redfern found government documents through the Freedom of Information Act (FOIA) that showed that government agents were themselves asking who these beings were. So in the beginning when this phenomenon began, they could not have been government agents. Thus the premise of the movie was just a vacuous Hollywood script.

In the second chapter of Redfern's book, he analyses several theories as to who the MIB are and what they are doing here on Earth. He lists hallucinations, hoaxes, tulpas and vampires, tricksters, civilian investigators, G-Men, time travelers, and demons and the occult as possible explanations. Each one seemed plausible, but how to decide?

Then it occurred to me that I could make contact with the MIB and ask them directly all the questions that needed answering. Go to the source, I thought. I could use my remote viewing capabilities and establish contact with the MIB and have them tell me why they are contacting witnesses who have seen UFOs. Why are they dressed in black? How do they appear and disappear suddenly? How can they walk through walls? What is their method of operation? Are they naturally evolved beings, and where do they come from? Are they here to harm or help us?

By talking with the MIB, many concepts that have been misunderstood by humanity are now clear and understandable.

[I received these] greetings from the MIB: "Dear Human Beings from the MIB Project Manager for Earth: The MIB are here on your Earth with peaceful purposes. We are a galactic race of beings who provide protection from harmful spiritual energy beings. Due to your physio-energetic nature, humans are susceptible to invasion by these unwanted energy beings. They want your energy and body. We are here to see that this invasion is stopped before any harm can come to you. This book explains in more detail our method of operation. Thank you, over and out. MIB Project Manager for Earth."

TIME TRAVEL: THE SCIENCE AND SCIENCE FICTION

The Manager showed me the individual letters of his name, and I drew them out on a piece of paper. He said that our Greek alphabet is close to their alphabet. Looking at the third letter *xi*, my drawing shows that that letter should be reversed. It should be a mirror image in the horizontal axis with the spaces facing left. Otherwise, that is his name. From my conversations with him, I was able to construct [a] stick-figure diagram that clearly shows what their purpose is here on Earth.

Looking at the top left of the diagram, there is an alien spacecraft (UFO) emerging *out of a wormhole* in the sky [italics mine]. The wormhole connects another dimension with ours. It enables the aliens to move between dimensions. You can think of it as a doorway. The opening of the wormhole is filled with low-density white hyperspace mist that makes the wormhole look like a regular cloud.

In fact, I had some man call me who said that he saw a UFO go into a cloud and it never came back out. He claimed that the UFO was hiding in the cloud! So I explained that *what he really was seeing was the opening of a traversable wormhole* [italics mine] with a low speed of light hyperspace energy flowing out of the other dimension into our dimension. The UFO was long gone into another dimension.

Just below the alien spacecraft is a person who is witnessing the appearance of the spacecraft. The person is so excited emotionally about this sighting that he just has to tell all his friends, put it on Facebook, write a blog or a book and so forth. Instead of just one person knowing about the spacecraft, tens of millions of people are now aware of it, as shown by the three stick figures. According to the MIB, this combined psychic

"What he really was seeing was the opening of a traversable wormhole."

energy keeps the wormhole open between dimensions. So the stick figure with the dark glasses, located just below the witness, is the MIB who intervenes in order to hush up the witness. Because the MIB are here to help us, they do not directly threaten the witness. They ask the witness about his sighting, getting all the details and recommending that he forget about it. In the conversation, they then throw in this non sequitur about how brakes on cars can sometimes malfunction. The human mind then takes this as a mildly disguised threat if the witness were to reveal something. The situation is puzzling because how could a simple sighting evoke such a threatening reaction?

After hushing the witness, *the MIB walked back out of dimension through the wormhole from which they had appeared* [italics mine]. If the witness follows them, their sudden disappearance is also baffling. The MIB Project Manager said that they wear black clothes because it makes them difficult to see when traversing the wormhole from our space to hyperspace. There is a light in the co-dimension that shines through the opening so the MIB can see the entrance. Many witnesses erroneously think that this light is emanating from a landed UFO. Even though the MIB wear a white shirt, when they turn around all you see is black. From my personal experience, a figure walking between dimensions is only seen as a black silhouette. The human eye can only see a short distance into this hyperspace region. In one case, I had to tell the being when I was able to see her as she walked between dimensions. When I shouted out that I could see her, she knew how close she could be but still remain invisible while looking into my room. In a case where the spacecraft enters unobserved, then the wormhole will dissipate and close naturally.

"It is when the wormhole remains open, due to the activity of the witnesses, that problems start happening."

It is when the wormhole remains open, due to the activity of the witnesses, that problems start happening. In other dimensions there are energy beings who can come into this dimension and cause us spiritual harm. These entities could be demonic energy entities of erratic frequency and low energy.... [Once] I encountered a foot-long rod creature that sucked out the hyperspace energy from my 4th green heart chakra energy vortex. The next day I developed chronic fatigue syndrome (CFS), a well-known medical condition.

On the left of the diagram is a second arrow coming out of the wormhole showing a wavy stick figure representing this dangerous energy entity mentioned by the Manager in his greeting. This entity can take possession of your body while suppressing your own energy. The slanted arrows point to two cases of possession involving a group of truck drivers, and the daughter of a woman who is well known in alternative medicine and energy healing.

The last reference is a fascinating experience with none other than the biblical demonic being called Beelzebub. Beelzebub killed all the tropical fish in an indoor pond by poisoning the fish flakes a friend of mine was feeding them. Fortunately I had developed a means of ascending demonic beings into angelic beings by boosting their energy and converting them to a single frequency. Notice the rings and hollow sphere that are used to ascend the demonic entity into an angelic being with two wings.... You never know when you might need this important technique in spiritual protection. I mean, the police officer who was attacked by Beelzebub committed suicide the weekend before he attacked me. So the MIB are here to provide protection from these types of unholy situations.

But of course, some entities slip through. I remember reading about a little boy who was playing with a friend of his. The boy started playing with his father's gun, and shot his friend dead. The little boy said, "I didn't pull the trigger." My explanation for this is that the entity wanted to create emotional turmoil in the people involved in order to boost its energy and possibly harvest the friend's energy field. So the entity pulled the trigger. Our judicial system needs to recognize what is really going on in these cases.

The MIB Manager told me that their planet is located in this dimension within our galaxy. They operate on many planets whose inhabitants require protection. For example, a whole village might see the UFO and keep talking about it. The MIB have to go in and hush up the villagers before word gets out to other villages. In my initial drawing of the diagram, I only had a straight line from the MIB planet into hyperspace.... The Manager observed the printout and noticed that there should be a wormhole into hyperspace at that point. I thanked him for noticing it and made the correction. He said they teleport through hyperspace using spacecraft that are designed by another alien race.

> *"They are not time travelers from the far future, but travel within a limited local time range. In this way they can obtain new, but older-model cars from decades ago."*

He went on to say that the MIB are naturally evolved beings. Their planet is similar to that of Earth except that it has a 50/50 water/land ratio. They are not time travelers from the far future, but travel within a limited local time range. In this way they can obtain new, but older-model cars from decades ago.

At the end of the conversation, he said goodnight and I asked him if he was finished working for the day. He said in an annoyed tone that, no, he still had to deal with all these bureaucratic papers such as MIB agent reports,

TIME TRAVEL: THE SCIENCE AND SCIENCE FICTION

project status and accounting that had to be sent to the home planet. He commented it was much more fun to roam the galaxy and explore new things. I guess that things don't change that much after all! MIB—Men in Bureaucracy.

But the question still remains as to how the MIB generate the wormhole in the first place? How do their spacecraft achieve lift without rockets? One of the clues can be seen in Redfern's book *The Real Men in Black*. One UFO investigator spotted a MIB and followed him down a corridor. The MIB then turned the corner down another hallway. When the investigator turned the corner, he found that the MIB had just disappeared. The length of the hallway was such that the MIB should still have been able to be seen, but that wasn't the case. The MIB was gone. The investigator continued walking along the hallway and came upon an intense electrical field that sent electrons running up and down his body.

So here is the clue that indicates how they create the wormhole: It turns out it can be understood from Einstein's Theory of General Relativity. This understanding leads into many hyperspace devices and capabilities, shown on the right side of the diagram, such as spacecraft propulsion, teleportation, wormhole generation, orbital debris removal, levitation and energy healing [italics mine]. So there is more exciting information in store for the reader as we explore the realm of hyperspace physics, who we really are, and answer all the questions that have puzzled us down through the ages.

In Gray Barker's book *Men in Black: The Secret Terror* there is supposedly a photograph on page 139 of Indrid Cold. The caption states that this notorious figure could be the head of the MIB! The caption states that this possible photograph of his is unlabelled and unattributed, possibly out of fear. So is this a photograph of him or not? And is he a notorious person?

Now St. Clair gets to the matter of Indrid Cold. He says:

I happen to have made contact with Indrid Cold. In the Ohio Valley he landed his spacecraft in the middle of the road, blocking the passage of an approaching car. Indrid then got out and walked slowly toward the man in the car. His arms were folded in front of him and he sported a large smile on his face. I asked Indrid about why he did this, and here is what he said. He folded his arms because he did not want the man in the car to think that he was carrying a weapon. He was smiling just to indicate that there was nothing to worry about and that he was friendly. It was to show that it was a friendly contact, nothing threatening.

*"This combination of magnetic field and
electromagnetic field is what creates a space-time
curvature to generate lift."*

I ran the permutation function in mathematics and found that his name spells Coil in DDD or Coil in 3-D with the initial R. He confirmed that this was the correct permutation. His spacecraft was described as a 3-D funnel chimney lantern. The base is a sphere on the top of which sits the funnel. The funnel is actually a tapered wire-wound solenoid that creates a magnetic field with a vertical gradient. The bottom is a spherical antenna that emits a wave traveling through the funnel. This combination of magnetic field and electromagnetic field is *what creates a space-time curvature to generate lift* [italics mine]. Not shown is the tripod landing gear. When I initially did the analysis of this vehicle, I did not understand how it generated lift. Then one day I found myself for some reason winding coils all day long. Then I was looking at the equations when Indrid started talking to me. He gave me some clues but still I did not understand. All of a sudden, a brilliant white flash of light hit me and I then I finally understood how his spacecraft worked. He was laughing his head off at seeing me get hit. I started laughing as well.

Afterwards, he told me that he was an explorer for his planet in this galaxy, sending information back to scientists who were analyzing the data. So this was the reason he was near Point Pleasant, West Virginia. A large wormhole had opened up and demonic entities started coming through. So he, like the MIB, was studying the phenomenon.

CHAPTER 21

ALIEN ROBOTS AND TIME MANEUVERING

It was only a little more than a week or so after Kenneth Arnold's now-famous June 24, 1947, encounter of the flying saucer kind—at Mount Rainier, in Washington State's Cascade Mountains—that a highly unusual aerial vehicle plunged to Earth on remote farmland of Lincoln County, New Mexico, not too far from the town of Roswell. The deeply controversial event has been the subject of dozens of books, official studies undertaken by the General Accounting Office and the U.S. Air Force, a plethora of television documentaries, a movie, and considerable media scrutiny and public interest. The admittedly weird affair has left in its wake a near mountain of theories to explain the event, including a weather balloon, a Project Mogul balloon secretly monitoring for Soviet atomic-bomb tests, an extraterrestrial spacecraft, some dark and dubious high-altitude exposure experiment, an atomic mishap, the crash of a Nazi rocket with monkeys on board, and an accident involving an early "Flying Wing"–style aircraft, built by transplanted German scientists who had relocated to the United States following the end of World War II.

It is no secret that I am distinctly skeptical of the idea that aliens met their deaths in the desert on that long-gone day in July 1947. And I consider that should we one day uncover the true story of what really occurred outside of Roswell, it will likely be one of secret military experimentation born out of the early years of Cold War shenanigans. But, of course, I may well be 100 percent wrong in my suspicions.

What if, incredibly, their point of origin was a far-flung
future of a distinctly human nature?

Keeping in mind the sentence immediately above, what if the Roswell affair is explainable in a very different, and wildly alternative, fashion? What if the weird craft and its strange crew were not the denizens of another galaxy, or even of the military of the immediate post–World War II era, after all? What if, incredibly, their point of origin was *a far-flung future of a distinctly human nature?*

Such a scenario may sound extreme and incredible to many, even to those who are of the opinion that something truly anomalous occurred in Lincoln County all those years ago. Nonetheless, such theories have been both expressed and endorsed. One of those who revealed his thoughts on this particular scenario was Lieutenant Colonel Philip Corso, coauthor with William Birnes of the much-debated 1997 book *The Day after Roswell.*

The sensational, and also deeply questioned and criticized, story told of Corso's alleged personal knowledge of the Roswell affair while serving with the military and of the way he allegedly helped to advance the United States, both scientifically and militarily, by secretly feeding certain fantastic technologies found in the craft recovered at Roswell to U.S.-based private industries and defense contractors.

Despite the fact that many have championed Corso as a solid proponent of the idea that extraterrestrials plunged to Earth in New Mexico in 1947, in reality, Corso was willing to consider something very different.

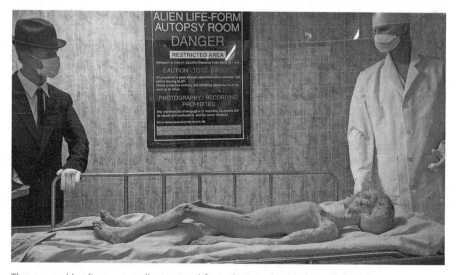

The unusual bodies reportedly removed from the wreckage of a craft found near Roswell in 1947, according to Lt. Col. Philip Corso, may have been genetically created beings from the future designed to withstand the rigors of space flight and time travel.

TIME TRAVEL: THE SCIENCE AND SCIENCE FICTION

The unusual bodies found within the wreckage of the craft, Corso maintained, were genetically created beings designed to withstand the rigors of space flight, but they were not the actual creators of the UFO itself. Right up to the time of his death in 1998, Corso speculated on the distinct possibility that the U.S. government might still have no real idea of who constructed the craft or who genetically engineered the bodies found aboard or in the vicinity of the wreckage.

Notably, Corso gave much consideration to the idea that the Roswell UFO was a form of time machine, possibly even one designed and built by the denizens of an Earth of the distant future rather than by the people of a faraway solar system.

> *Corso gave much consideration to the idea that the Roswell UFO was a form of time machine, possibly even one designed and built by the denizens of an Earth of the distant future.*

It's vital that I make it clear that Corso's story has been the subject of a great deal of intense debate. His account relative to Roswell has been both vocally championed and loudly denounced. Many seem unsure what to make of it all. But, unfortunately for those trying to make some sense of the situation, ufology has always been like that when it comes to matters of a highly volatile nature. We never get a definitive answer. It always ends up, to use a terrible but appropriate pun, in a "grey" area.

If history shows there's nothing to the story of Philip Corso, then so be it. But if there is even a small nugget of truth to the story, then here is something we should muse upon: Maybe, by studying the Roswell materials, officialdom has learned something deeply troubling and terrible about our future—something it dares not share with us, the populace at large.

Is this, perhaps, the reason why the Roswell affair is one still shrouded in overwhelming secrecy more than 60 years after it occurred? To paraphrase *The X-Files*, when it comes to UFOs and Roswell, "the truth" may not be "out there" after all. Instead, it might be countless millennia ahead of us.

The fact that speculation exists that both the Men in Black and the so-called Grey aliens may really be traveling, biological robots is certainly intriguing. In the summer of 1997, one of the most controversial UFO-themed books ever written was published. Its title: *The Day after Roswell*. Ghosted by William Birnes, the editor of the now-defunct *UFO Magazine* and of the History Channel's also defunct *UFO Hunters*, the book tells the story of one Lieutenant Colonel Philip Corso of the U.S. Army. Corso's story was both amazing and groundbreaking. But was it true? While some in the field of ufology embraced the story, many certainly did not, preferring instead to view the book as either government disinformation, intended to confuse the truth of what really happened at Roswell back in the summer of 1947, or nothing more than an elaborate and ingenious hoax designed to make money from the gullible and the "I want to believe" crowd.

According to Corso, he spearheaded, nearly singlehandedly, a secret program designed to seed alleged alien technology and wreckage—recovered from the Foster Ranch in Lincoln County, New Mexico, by the U.S. Army Air Force's 509th Bomb Wing in July 1947—into the private sector. Incredibly, as time went along, Corso came to believe that the Greys were actually time travelers and that the Roswell craft was really a time machine.

As a result of this strange and sensational clandestine operation, so Corso maintained, the United States was soon able to understand and even back-engineer at least some of the extraterrestrial materials. Fiber optics, transistors, night-vision equipment, and computer chips were all, allegedly, a direct outgrowth of the extensive studies of the Roswell materials. But it's not so much the technology reportedly found at Roswell that we need to focus our attention on. Rather, it's the bodies of the beings allegedly found strewn around the crash site.

Contrary to what the UFO community said or assumed, Corso never explicitly stated that the Roswell corpses were extraterrestrial. In fact, what he really said was

One theory posits that such technology as night-vision equipment (pictured here), fiber optics, transistors, and computer chips were all back-engineered from extensive studies of the Roswell materials.

TIME TRAVEL: THE SCIENCE AND SCIENCE FICTION

quite the opposite. In Corso's story, the so-called Grey aliens are actually created, built, or grown to perform specific tasks. As for the creators of the Greys, it was they, Corso maintained, who were the real aliens. Corso also said that as far as he was aware, no one has ever seen the real aliens, only their black-eyed, large-headed worker drones.

William Birnes said that Corso described the Greys as "an android or biological robot. He said it had no digestive system whatsoever and was connected electronically to the navigation controls of the spacecraft."

Corso himself used the term "extraterrestrial biological entities," or EBEs, and stated: "Perhaps we should consider the EBEs as described in the medical autopsy reports humanoid robots rather than lifeforms, specifically engineered for long distance travel through space or time."

He expanded on this, outlining the profoundly weird nature of the Greys: "While doctors couldn't figure out how the entities' essential body chemistry worked, they determined that they contained no new basic elements. However, the reports that I had suggested new combinations of organic compounds that required much more evaluation before doctors could form any opinions. Of specific interest was the fluid that served as blood but also seemed to regulate bodily functions in much the same way glandular secretions do for the human body. In these biological entities, the blood system and lymphatic systems seem to have been combined. And if an exchange of nutrients and waste occurred within their systems, that exchange could have only taken place through the creature's skin or the outer protective covering they wore because there were no digestive or waste systems."

A similar scenario has been offered by a group called the Hybrids Project, which has stated: "These small grey beings are in fact biological workers. They appear to be fabricated workers which are the result of extremely advanced genetic manipulation. The Greys have succeeded in creating autonomous creatures, what we might call androids, who carry out mundane or dangerous tasks for the Tall Greys."

Today, more than two decades after it surfaced, the strange story of Colonel Corso continues to divide the UFO research field. There is, however, a very good reason why I bring it up in the pages of this book. The idea that the Greys are not born as such but are grown or created has attracted the attention of a man named Nigel Kerner. He has developed an extensive and controversial theory that has a direct bearing upon matters relative to both alien and human immortality. It's a theory detailed in Kerner's books *The Song of the Greys* and *Grey Aliens and the Harvesting of Souls*—the latter being a book for which I provided a cover blurb.

Danielle Silverman undertook a great deal of research for Kerner, and in 2011 she sent me a very illuminating statement for publication. She told me the following, which makes for fascinating reading. The immortal human soul, she says, may well be "a derivational information field that comes out of a natural cadence that came into the Universe with the big bang. This field holds the power to maintain information in what [Kerner] called a morphogenetic electro-spatial field with an eternal scope of existence in whatever form circumstance allows."

Silverman suggested that the soul may actually be a form of mechanism for storing data. Keeping that scenario in mind, she noted that Kerner is of the view that if the Greys are, indeed, some kind of biological robot, then the soul may amount to "an analogue" of the very things that manufactured them in the first place.

As Silverman also told me, Kerner suspects that the Greys are, in essence, biological machines: they are entities created and possessed of high degrees of intelligence that are dispatched throughout the universe, effectively on exploratory missions—something that sounds like the Men in Black. Kerner is keen to stress, however, that the Greys are not supermen—or, rather, super-aliens. Rather, he believes them to be prone to their own equivalents of the wear and tear that affect all of us throughout our lives. The result is that the Greys are not invulnerable. And, given their mortality, they are driven to seek out, to understand, and even to try to find a way for themselves to have immortal souls.

Author Nigel Kerner speculates that "the Greys" are biological machines: entities created and possessed of high degrees of intelligence that are dispatched throughout the universe, effectively on exploratory missions.

Silverman explained to me that this latter issue gets right to the heart of the agenda of the Greys. Fully aware of the fact that they are not invulnerable to the rigors of time, the Greys are desperate to find a way to survive—and to survive forever. While the Greys may, at the very least, carry certain DNA coding from the unknown extraterrestrial entities that created them in the first place, this does not allow them to survive eternally. But, via highly advanced genetic manipulation and exploitation of the human race, the Greys may manage to secure some kind of unending life—hence the near-endless number of reports of alien abduction and the reaping of bodily cells, DNA, sperm, and eggs that so many victims of such abductions report. Silverman suggests that the dark agenda of the Greys cannot work— that it's impossible for them to essentially create souls for themselves. But they appear not to be aware of that fact, which is why they endlessly attempt to do exactly that, and with no end in sight, as far as we can tell.

Perhaps the most fantastic revelation that Silverman shared with me on this matter concerned Kerner's conclusion that the most profitable way to traverse the cosmos is via what he has termed Fields of Death. It's not as ominous as it sounds. Basically, Kerner has concluded that there is what we might call a zero-point area that exists in the specific spaces between atoms. This allows for incredible distances to be traveled in extraordinarily short periods of time—even allowing us to reincarnate on other worlds, in other solar systems, and maybe even in faraway galaxies. The Greys, seemingly, are aware of this post-death ability that we all apparently possess. In much the same way they are trying to perfect the ultimate soul to keep themselves alive, the Greys are doing their utmost to understand this afterlife ability to traverse the universe at incredible speeds—and to replicate this, too.

As for Kerner himself, he provides us with the following: "I contend that there is something hugely meaningful about humanity, something virtual reality can never match, something artificial intelligence can never be programmed with. Words such as conscience, compassion, warmth, kindness, generosity, spontaneity, imagination, inspiration and creativity, give some hint as to what this something might be."

In terms of the specific nature of this mysterious something, Kerner believes it to be the soul. Kerner also concludes that the Greys are desperate to understand the nature of the soul, to the extent that they may even be able to replicate it in themselves genetically. Doing so is the only way in which they can achieve unending existences, says Kerner. On this issue, he provides us with this:

"It is my thesis that the Greys, in contrast, are purely physical creations and thus completely subject to the entropic momentums that break down and decay physical states. They have no line of connection to any non-physical state that might lie beyond the physical mass-soaked materiality of this universe, no 'soul'. Without this component, the Greys are completely subject to the breakdown momentums implicit within a physical universe. In my books I document alien genetic engineering at DNA level to evidence my theory these entities are attempting to 'piggy back' our facility as human beings for eternal existence, hence their apparent fascination with the human reproductive capacity."

TIME TRAVEL: THE SCIENCE AND SCIENCE FICTION

This is something echoed by an alien abductee named Allison Reed, presented by David M. Jacobs in his book *The Threat: Revealing the Secret Alien Agenda*. In an interaction with a Grey she learned something incredible. Reportedly, the Grey in question "claims that he and his grey people are the result of genetic manipulation that some higher species, I guess, played God and mixed and matched and whatever. He and his people were created through a genetic alteration through a higher intelligence. I don't know what they were created for. But my understanding is that they were created for a purpose and, through the years, they weren't able to reproduce themselves anymore. From what he told me they didn't start this. They were a result, just like the hybrids are, from something else. From a higher intelligence."

And now, the presently soulless Greys are on a mission to ensure that immortality is one day theirs. But only by understanding the nature of the human soul, figuring out what it is that gives us a soul-based immortality, and using advanced science and technology can they hope to achieve the one thing they so crave: never-ending life. Perhaps that's why the Greys view us not as humans, as people, or even as a species but as containers. It's not our physical bodies that are the most important things to them. It's that ethereal, near-magical thing which is within all of us that they both want and need: the human soul, the key to living for all eternity.

CHAPTER 22

A CROP CIRCLE CONNECTION

Crop circles constitute one of the most fascinating phenomena of latter-day times. The theories for who or what is responsible for them include aliens, beings from other dimensions, military experiments, and Earth itself warning us of our reckless, dangerous ways. And, of course, time travelers enter the picture, too. You knew that was coming. If you look carefully at the available data, you'll see that there's no doubt about the culprits. It's us: the human race. While some consider the circles to be artwork of the kind that allows the makers to commune with higher, supernatural entities, others view the makers as nothing but hoaxers. Still others believe that time travelers are leaving messages in the circles suggesting we need to reel in our warlike ways. We'll start investigating this thread with a man named Matthew Williams.

> *Still others believe that time travelers are leaving messages in the circles suggesting we need to reel in our warlike ways.*

A former special investigator with the British government's Customs and Excise agency, Williams holds the distinction of being one of the few people in the world arrested, charged, and convicted for making a crop circle. The actual crime was that of causing damage to the field in which the formation was made. I will be the first to admit that Matthew is a good friend of mine, and a great deal has been written about his reasons for making crop formations that is either misleading or downright untrue. More importantly, the notion that people such as Williams can be dismissed as mere hoaxers is an utter failure to understand and appreciate the philosophy and experiences of the human circle makers.

In an extensive interview I conducted with Williams, he began: "In 1992, when I got involved in crop circle research, there were a lot of people who believed that crop circles could only be made by otherworldly forces. Admittedly, I didn't know very much about crop circles back then, and hearing some of the arguments of 'bent nodes' and 'blown expulsion cavities,' coupled with UFO sightings, made the whole subject seem extraordinary. Without researching the subject for myself, I was inclined to believe the so-called experts and their seemingly plausible theories about aliens making crop circles.

"My early research efforts involved investigating areas where crop circles were expected to appear. I coordinated efforts with other friends and researchers via radios, and we patrolled fields in our cars, armed with night-vision equipment. Back in those hazy days, I used to think that people who made crop circles were trying to confuse us and lead us away from the path of discovering deeply encoded alien messages. I was on a mission to stop man-made crop circles appearing."

Williams explained to me the motivation behind his mission: "Many researchers I knew believed that the crop circles were alien in origin and felt there was nothing to be gained in trying to police human circle makers, as, in the opinion of the researchers, there were very few crop circles created by people anyway; and the re-

The theories for who or what is responsible for crop circles include aliens, beings from other dimensions, military experiments, Earth itself—and, of course, time travelers.

TIME TRAVEL: THE SCIENCE AND SCIENCE FICTION

searchers I knew were sure that they could tell the difference between a man-made crop circle and an alien-made crop circle.

"The standard method of researching crop circles generally involved going into them on the same day they had formed and studying what was present. My own motivation was to get closer to whatever was creating the crop circles, at night and when the circles were formed. If we had caught any human in the act of making crop circles, then the plan was to scare them off or perhaps even call the police. In fact, our field surveillance activities were actually welcomed by the police as a form of anti-criminal vigilance. If we could be sure we had cleared a field of any human-made crop circles, then logic dictated that if a crop circle did appear, it must be genuine."

Williams admitted, however, that these plans were both "rather naive" and "fruitless" with "none of us even getting close to a crop circle being created under weird or 100 per cent confirmed fields-clear-of-hoaxers."

He elaborated: "It wasn't until about 1994 that I decided to seek out the human crop circle makers. What subsequently happened could be termed the 'other side' of the story. By this time, I had a good idea that many crop circles were made by humans, but was unsure if there was a level of disinformation being spread. It was then I met [crop circle researcher, maker, and personality] Rob Irving and heard his immortal quote, which I still use to illustrate the simplicity of the crop circle argument: 'You'll never be sure until you actually make one.'

> *"The crop circle makers I spoke to did not debunk the theories that paranormal events take place in crop circles, even if they are man-made."*

"What I learned next was totally unexpected, as the crop circle makers I spoke to did not debunk the theories that paranormal events take place in crop circles, even if they are man-made. Indeed, there was much talk of strange lights and events happening to the crop circle makers while they made crop circles at night.

"I was given countless tales of UFO activity and even strange synchronicities which took place in the man-made crop circles, and which made the crop circle makers wonder if they were being influenced by a higher, guiding hand—perhaps at the design stage, or out in the field.

"I befriended a number of circles teams and watched crop circles being made under various conditions. Back in the early days, I thought that speaking to the human crop circle makers would help me to determine which crop circles were made by unknown persons or paranormal forces. Later on, however, I became convinced that human-made crop circles made up the whole sum of the complex pictograms, leaving only very simple crop circles as possible unknowns."

TIME TRAVEL: THE SCIENCE AND SCIENCE FICTION

Williams makes an important point that whether you agree with him or not, this takes him just about as far away from being a mere hoaxer as it is possible to get: "I believe that the human crop circle makers may be guided by a connection to a higher intelligence, a gestalt or group mind; and the symbols that are put down are deep-routed subconscious expressions. This would explain why some telepaths or psychics claim to have seen crop circle designs before they appear—because we are all dialed into the same group-mind 'internet.'"

On this same matter, he continues: "The synchronistic events which happen each year in the unfolding designs that the crop circle makers come out with are often remarked as being too much of a coincidence. Crop circle designs have been repeated in different parts of the country by teams who didn't share designs beforehand. There have even been teams who have gone out on the same night to the same field [and] made designs, not knowing the other team was present, and their designs have had similarities. The chances of this happening, without planning, are slim; so it would appear that something paranormal is working for the crop circle makers."

Williams asks: "Why do crop circles have power? I know that crop is a living medium. Unlike any other type of canvas that an artist may use, crop has a life-force energy which, when used on such a large scale, makes crops blend natural Earth-energy magic with human-energy magic. What crop circle makers may have redis-covered is the ancient art of natural magic—working with large emblems as part of the energy system of the Earth.

"I believe that the human crop circle makers may be guided by a connection to a higher intelligence, a gestalt or group mind," observed Matthew Williams, a former special investigator in the United Kingdom.

"Take sacred sites such as stone circles: these were erected as temples for people to visit, perhaps as part of a sacred quest seeking answers. Today, crop circles fill a similar role, attracting people who ask questions and putting visitors closer to paranormal sources."

On this theme, Williams adds: "Many people have expressed concern to me about humans making crop circles and trying to instill magical energy in them. Along with this comes a distrust of the motives of crop circle makers. However, if researchers are to be believed on the healing energies of many crop circles, then this is not a negative thing, by definition."

"Take sacred sites such as stone circles: these were erected as temples for people to visit, perhaps as part of a sacred quest seeking answers. Today, crop circles fill a similar role."

As for the process of creating crop circles, Matthew explains: "Over the years from 1994, I participated as crop circle maker, designer of crop circles, and observer to other persons creating crop circles. They are typically created in the dark, just after it gets too dark to be seen in the field. Typically, this was from 10:00 to 11:00 p.m. in the British summer months. Teams of between two and nine would work under the guidance of whichever team member designed or studied the design. Coordination is essential among team members.

"The actual laying down of the crop circle is usually done in stages. The first stage is known as foot-lining, which means the creation of the thin part of the design, without doing the shaded areas of the design, which represent flattened areas of crop. One person will hold one end of a tape measure. The distance on the tape measure is found and the other end of the tape is held. One end of the tape is held in place and the other end is walked around, creating your circles. Holding the tape tight is important, as over a distance of 100 feet you can get considerable sag in the tape which can, if not handled correctly, cause circles to have wobbly edges.

"To etch in the internal designs, two ends of the tape are held, which gives a straight line, and a third person walks along the edge of the tape. It is important to note that these etching lines, created by the width of your foot in the crop, are often visible as thin lines hinted at when aerial photos of crop circles are zoomed in on. Marker posts are used to guide tape-measurement points. These points are driven into the ground and are often made of bamboo posts, similar to those found in garden use.

"Once the design is fully etched with foot-lines, then the process of filling takes place. The teams have to be briefed as to which parts are to be filled or left standing. There are usually simple rules to follow to make sure you don't flatten the wrong area. Often tramlines or tractor lines are mistaken for starting or stopping points for filling work. In the filling stage, planks of wood affixed with ropes are used to press down with large areas of crop. Approximately one meter in width, these boards will leave tell-tale signs of human construction when viewed from the ground or the air.

"The appearance of many circles will be a combed effect. Some types of crop, brittle plants or late-in-the-season plants, fall down well and do not grow back up to the light. Formations created in these fields look very impressive because the ground-lay, with its flat uniformity, shines from the air when photographed from the correct angle."

Williams makes a valuable observation: "Crop circle makers tread a thin line. On the one hand, they want people to come to them and get experiences from the crop circles. If crop circles were known as a 100 per cent man-made hoax, then the public might turn away. So this is why authorship of most crop circles remains anonymous, for legal reasons and for reasons of sensitivity. However, to stem the rot of blind belief in crop circles and to make sure crop circle makers and their motives are known to those who have delved deeper, there is a dialogue between some researchers and crop circle makers. This is to allow the more open-minded to see the bigger picture at work."

Again, whether people agree with Williams and those who follow a similar path, the fact is that in the minds of the people who create them, they are most assuredly not hoaxers, merely on a path to spoil everyone's fun and deceive them. No; if people such as Matthew are correct, then (a) at least some of the human crop circle makers seem to be guided by a form of higher, unidentified intelligence; (b) the work of the crop circle makers appeared to be inextricably linked to the worlds of ancient magic and archaic ritual; and (c) the public image of the "crop circle hoaxer"—which is often offered by, and to, the collective UFO and crop circle research communities—is about as far away from the truth as it is possible to get.

Now, it's time to address the matter of time travelers and crop circles.

Of the various theories for who or what might be responsible for the crop circles, *Live Science* writer Benjamin Radford notes that some observers, such as "molecular biologist Horace Drew, suggest that the answer lies instead in time travel or alien life. He theorizes that the patterns could be made by human time travelers from the distant future to help them navigate our planet. Drew, working on the assumption that the designs are intended as messages, believes he has decoded crop circle symbols and that they contain messages such as 'Believe,' 'There is good out there,' 'Beware the bearers of false gifts and their broken promises,' and 'We oppose deception' (all, presumably, in English)."

TIME TRAVEL: THE SCIENCE AND SCIENCE FICTION

"He theorizes that the patterns could be made by human time travelers from the distant future to help them navigate our planet."

Writing for the U.K.'s *Sun* newspaper, George Sandeman informed readers that Drew maintains that crop circles "contain hidden messages from aliens or human time travellers." Sandeman continued: "Dr. Horace Drew, 61, also suggested that the mysterious phenomena are a method extra-terrestrial beings are using to try to peacefully communicate with the human race. Dr. Drew, who holds a PhD in Chemistry from the prestigious California Institute of Technology, made the claims at the UFO & Paranormal Research Society of Australia last Wednesday.

"He revealed that he had a lifelong fascination with UFOs after seeing an un-identified, silver and windowless craft hovering near his home as a 10-year-old in Florida, USA. He moved to Australia in the 1980s and has been studying crop circles for 20 years.... He said that the study of the peculiar plant arrangements had led 'to at least one major breakthrough: the discovery of a more advanced binary code than our computers currently use.' Additionally, other crop circles 'show schematic images of the future for astronomical or human events.'"

Then there is "alien investigator" Scott C. Waring, as the U.K.'s *Express* news-paper described him. Writer Jon Austin of the *Express* said of Waring: "Mr. Waring has previously claimed the formations are the marks in the ground left by aliens who landed a UFO in the field. Crop circles are often found in agricultural areas in the west country of England. But, he has now expanded his far-fetched conspiracy theory to include the possibility they are messages sent back by future generations of earth-lings, and therefore proof that time travel is invented in the future. Crop circles have long been alleged to be caused by alien visitations. However, Mr. Waring is known for making statements of fact in connection with alleged UFO evidence based on his own speculation.

"Crop circles, which often appear in the south west of England, were ex-posed as elaborate hoaxes during the 1990s when many pranksters came for-ward to show how they made the often intricate designs using wooden planks. Despite this exposure, some alien chasers still believe they are left as signs by aliens or are even the landing marks of a UFO."

Molecular biologist Horace Drew maintains that crop circles, such as this one in Bavaria, Germany, "contain hidden messages from aliens or human time travellers."

TIME TRAVEL: THE SCIENCE AND SCIENCE FICTION

It's possible that the following story of crop circles has a tie-in with time travel. In a March 2009 article titled "Dog Walker Met UFO 'Alien' with Scandinavian Accent," journalist Sarah Knapton wrote in the pages of Britain's highly respected *Daily Telegraph* newspaper that the British Ministry of Defense had recently released into the public domain, via the terms of the nation's Freedom of Information Act, a number of formerly classified files on UFOs. One of those files, said Knapton, detailed the account of "a dog walker [who] claimed she met a man from another planet who said aliens were responsible for crop circles." But was the mysterious visitor really an alien? Or could he have been a time traveler? Yes, he most assuredly could have been from another time.

As the story went, the anonymous woman had telephoned the Royal Air Force base at Wattisham, Suffolk, in a state of considerable distress and with a remarkable tale to tell. As the woman told the operator at the base, the incident in question had occurred while she was walking her pet dog on a sports field close to her home near the city of Norwich at about 10:30 p.m. on the night of November 20, 1989. She had been approached by a man with a "Scandinavian-type accent" who was dressed in "a light-brown garment like a flying suit."

Royal Air Force documentation on the case notes the following: "He asked her if she was aware of stories about large circular flattened areas appearing in fields of wheat, and then went on to explain that he was from another planet similar to Earth, and that the circles had been caused by others like him who had traveled to Earth." The man assured the woman that the aliens were friendly but that "they were told not to have contact with humans for fear that they would be considered a threat."

It would be very easy for people from a faraway century to pass themselves off as extraterrestrials.

Quite understandably, the woman said she was "completely terrified" by the encounter. She added that after about ten minutes, the strange man left. But things were not quite over: as the woman ran for the safety of her home, she heard a "loud buzzing noise" behind her and turned around to see "a large, glowing, orange-white, spherical object rising vertically" from behind a group of trees.

The Royal Air Force operator who took the statement from the woman said the conversation lasted about an hour and described it as a "genuine call." While I am sure that the call was genuine, I find it hard to believe the woman was confronted by an alien. After all, the man could speak English and looked completely human. I strongly suggest there was a high degree of manipulation in this case. It would be very easy for people from a faraway century to pass themselves off as extraterrestrials. The next case could fall into that category, too.

This incident I'm about to share with you occurred in 1989 and involved a man named Paul Farrant and a Woman in Black. I met Paul in the United Kingdom back in the summer of 1997. Paul had spent 10 days checking out the crop circles that had appeared during that summer in the county of Wiltshire, England, which is where most of the formations are made on a yearly basis. I was there for two weeks with Irene Bott of the Staffordshire UFO Group and gave a few lectures in the area at the same time, which is how I got to know Paul. He had a story to tell that involved missing time, but not at all of the type associated with alien abductions. Paul told me that he had been walking around one of the formations just as the sun was starting to rise when a woman appeared, seemingly out of nowhere. She was dressed in a long black cape that had a black hood. Paul said her skin was extremely pale, and her face was very thin. She engaged him in what seemed to be a very short conversation, warning him not to dig too deeply into the subject.

Not only had Paul lost 45 minutes of his life, but his watch had gone forward three hours.

The Woman in Black then turned and exited the huge and intricate formation. For reasons that he was unable to understand, Paul felt both nauseous and frightened. Here's the strangest part of the story: Paul found himself quite confused because, although it seemed to have been a brief encounter, when he checked the time, he realized there was far more to it. According to his watch, close to three quarters of an hour were missing from Paul's memory. And although the woman had said something to Paul of a warning nature, he couldn't remember exactly what he said in response to her. It puzzled and worried him for weeks.

One final thing: not only had Paul lost 45 minutes of his life, but his watch had gone forward three hours. There's no doubt that this is a very strange story, but it demonstrates that when one is in crop circles, time-based anomalies are not unknown.

CHAPTER 23

MUTILATIONS IN THE FIELDS

ince at least 1967, reports have surfaced throughout the United States of animals, chiefly cattle, that have been slaughtered in bizarre fashion. Organs are taken, and significant amounts of blood are lost. In some cases, the limbs of the cattle are broken, suggesting they have been dropped to the ground from a significant height. Evidence of extreme heat, apparently used to slice into the skin of the animals, has been found at mutilation sites. Eyes are removed, tongues are sliced off, and, typically, the sexual organs are gone.

While the answers to the puzzle remain frustratingly outside of the public arena, theories abound as to the cause. They include extraterrestrials, engaged in nightmarish experimentation of the genetic kind; military programs involving the testing of new bio-warfare weapons; occult-based groups that sacrifice the cattle in ritualistic fashion; and government agencies secretly monitoring the food chain, fearful that something worse than "mad cow disease" may have infected the U.S. cattle herd—and possibly, as a result, the human population. Another theory for the mutilations is that they are the work of time travelers.

You will recall from the chapter "Timelines and the Roswell Incident" that Jim Penniston—formerly of the U.S. Air Force and one of the key military players in the famous UFO encounter at Rendlesham Forest, Suffolk, England, in December 1980—underwent hypnotic regression in 1994 as part of an attempt to recall deeply buried memories about what occurred to him during one of Britain's closest encounters. Interestingly, while under hypnosis Penniston stated that the presumed

Reports of cattle mutilations dating to the late 1960s have attributed the phenomenon to extraterrestrials aboard UFOs, military programs, occult groups, government agencies, and time travelers.

aliens are, in reality, visitors from a far-flung future. That future, Penniston added, is very dark, in infinitely deep trouble, and polluted, and the human race of the time is overwhelmingly blighted by reproductive problems. The answer to those massive problems, Penniston was told by the entities he met in the woods, was for the entities to travel into the distant past—to our present day—to secure sperm, eggs, and chromosomes in an effort to ensure the continuation of the severely waning human race of tomorrow.

There is a reason why I mention this again. I have a very similar account to share with you.

The story comes from a man named Dan Salter. He was the author of a book titled *Life with a Cosmos Clearance*. Salter told me that cattle mutilations were not the work of aliens but of time travelers. The scientists of the future were performing nightmarish tests on not just cattle but on people, too. Centuries from now, Salter explained, our gene pool will be all but shot, and to try to keep us alive, terrible experimentation will be undertaken on a massive scale. The processes will even involve blending animals and humans into creatures that would be hard for us to recognize as people, said Salter.

The scientists of the future were performing
nightmarish tests on not just cattle but on people, too.

TIME TRAVEL: THE SCIENCE AND SCIENCE FICTION

But before we get to the cattle-mutilation angle of this, let's first take a look at one particular case that may be proof of Salter's strange claims. It was in the late 1980s when ufologist Leonard Stringfield, highly active in ufology at the time, received an astonishing and terrible report concerning an incident that supposedly occurred in early 1972 in Tong Li Sap, Cambodia, situated in southeast Asia, while the Vietnam War was still raging. The story is told in Stringfield's *UFO Crash Retrievals: The Inner Sanctum*, which was self-published in July 1991. A group of expert marksmen were secretly and silently parachuted late one night into an area bordering North Vietnam. The operation was a vitally important one: to take out a North Vietnamese facility that, as U.S. intelligence deduced, was clandestinely listening in on top-secret conversations between high-ranking American personnel in South Vietnam.

The team camped down for the night, prepared to make a full assault on the North Vietnamese team as dawn broke. It never happened. Rather, it didn't happen the way that it had been envisaged. The group made a skillful and stealthy approach on the area, using the dense foliage for cover, but found to their terror that something unearthly had changed the situation drastically.

As the team got closer to the area in which the North Vietnamese unit was hunkered down, they suddenly found themselves confronted by a large, ball-like craft that sat atop three sturdy metallic legs. The craft suddenly began to hum, which caused instant sickness, dizziness, and disorientation on the part of the U.S. troops. Fighting the inclination to throw up and needing get out of the area quickly, they were suddenly rooted to the spot by a group of strange-looking humanoid creatures that today we would call "the Greys" of UFO lore. Barely believing what they were seeing, the group was even more horrified by what the creatures were doing: handling various human body parts and placing them into large containers: arms, legs, torsos, heads—the grisly list went on and on. Some were the remains of white people, others were black, and several looked like Vietnamese. Managing to keep the sickness and dizziness under some degree of control, the team crawled forward on their stomachs. The commanding officer silently gave the order to fire, and salvos of bullets slammed into the bodies of the creatures. Most seemed barely fazed by the assault, aside from one that was said to have been killed by a shot to the head. Several of the U.S. troops lost their lives, and others were left severely burned by the effects of an unknown weapon. The aliens then quickly retreated, loaded the containers into the craft, and vanished into the skies.

In no time at all, another team was on the scene—"CIA types," as one of the survivors told Stringfield. All of the surviving men were given certain mind-altering drugs to try to make them forget the incredible affair—which apparently worked, or so we're told. At least, for a while they worked. In the late 1980s, however, two members of the team started to experience dramatic and nightmarish flashbacks to those events of April 1972, which prompted one of the team to contact as many of the others as he could find. Two were dead, three could not be located, but the rest were able to meet up in August 1988, when they decided that the story needed telling.

The story is still not over. One of Stringfield's retired military sources quietly informed him that the story wasn't hidden just because of the grisly and terrifying

TIME TRAVEL: THE SCIENCE AND SCIENCE FICTION

nature of it but also because the military had learned, to their shock, that the aliens were actually biological robots sent back from the future to secure human body parts to try to create something more than human. Or less than human, depending on your perspective. True or not, this story is very similar to the scenario that Dan Salter presented related to cattle mutilations. While we don't have evidence, so far, of these hideous experiments on people in centuries ahead, we are able to vindicate the nature of the cattle mutilations—as you will now see.

With that, we come to the matter of the alleged time-traveling creatures said to live deep inside a certain place in New Mexico. This picturesque area, the New Mexican town of Dulce, located in the north of Rio Arriba County, is steeped in mystery. It's home to fewer than 3,000 people and has a square mileage of barely 13. Its origins date back to the nineteenth century. It's not what goes on in Dulce that concerns us here, though. Rather, it's what is said to be going on far *below* the town, in myriad tunnels, caverns, caves, and hollowed-out chambers, where untold numbers of dangerous and hostile aliens are said to live. Even worse, the U.S. government has had the fear of God (or of the aliens) put in them to such an extent that they dare not descend into that deadly, dark realm far below Dulce's huge Archuleta Mesa.

Today, tales of underground bases, in which nefarious experimentation is widespread, are all over the internet. Just type "Underground Base + UFOs" and you'll find

Far below Dulce, New Mexico, are myriad tunnels, caverns, caves, and hollowed-out chambers, where untold numbers of dangerous and hostile aliens are said to live.

an endless array of tales of the controversial kind; they are overflowing with paranoia and tales of menace. Such tales were far less known and shared in the 1970s, which is when the Dulce stories began to surface, specifically in the latter part of the decade. What makes the Dulce story so notable is that the initial rumors about the vast, alien facility miles below ground level came not from wide-eyed conspiracy theorists but from a number of people who worked deeply in the clandestine worlds of counterintelligence and disinformation. The latter is described as "false information deliberately and often covertly spread (as by the planting of rumors) in order to influence public opinion or obscure the truth," while counterintelligence is defined as "organized activity of an intelligence service designed to block an enemy's sources of information, to deceive the enemy, to prevent sabotage, and to gather political and military information."

In other words, we're talking about spies, secret agents, lies that might be truths, and truths that might be lies.

As for the Dulce story, it suggests that there was a violent, deadly altercation in what became known as the Dulce base at some point in 1979, at which point the U.S. military, along with numerous scientists and engineers, were forced to flee for their lives. What had begun as a fairly amicable arrangement between the aliens (of the black-eyed "Grey" type) and the government team was now over. Irreversibly so. The Dulce base was now in the hands of a band of extraterrestrials who were done with the human race.

Back in the 1970s, Paul Bennewitz—who died in 2003 in Albuquerque, New Mexico—had his own company that stood adjacent to Kirtland Air Force Base. Its name was Thunder Scientific. All was good, as Bennewitz had a number of good contracts with the military. And living and working so close to the base made things comfortable and handy for Bennewitz. It was the perfect relationship. Until, that is, it wasn't. In shockingly quick time, Bennewitz's life began to fragment in chaotic fashion. How and why did such a thing happen?

It's important to note that by the late 1970s, Bennewitz had been interested in UFOs not just for years but for decades. He had a large library of books on the subject and subscribed to a number of related newsletters and magazines. On occasion, late at night and in the early hours of the morning, Bennewitz had seen strange, unidentified objects flying over Kirtland Air Force Base and the nearby, huge Manzano Mountains. They could have been early dronelike craft being tested secretly. For Bennewitz, they were alien craft.

Bennewitz's head spun. He came to believe that aliens were secretly in league with the U.S. Air Force and that much of the clandestine program was run out of Kirtland. He shared his views with staff at Kirtland, the Central Intelligence Agency, the National Security Agency, the Defense Intelligence Agency, the Pentagon, his senator, his congressperson, and just about anyone and everyone in a position of power and influence. It was all but inevitable that as a result of the lengthy letters he fired off about a secret alien–human operation at Kirtland, someone would take notice. Some did. While one school of thought suggests that Bennewitz was indeed tracking

New Mexico resident Paul Bennewitz saw strange, unidentified objects flying over Kirtland Air Force Base and the nearby Manzano Mountains.

the movements of UFOs in the skies over Kirtland, another suggests that Bennewitz had actually stumbled on test flights of new and radical aircraft, such as the afore-mentioned drones. Hewing to the latter scenario, the government (as a collective term for all of those agencies and individuals that Bennewitz approached) decided first to politely and quietly request that Bennewitz bring his research to a halt. This was like a red rag to a bull. Bennewitz would hear none of it. He was primed and ready to go after the U.S. government and confirm what he saw as the dark and sinister truth of Uncle Sam's liaisons with aliens. One man against the government? It was clear who was going to win, although Bennewitz couldn't envisage such a thing at all.

In ingenious fashion—but, from the perspective of Bennewitz, in terrible fash-ion—a plot was initiated to, in essence, give Bennewitz exactly what he wanted to hear. To accomplish this, well-placed government agents, intelligence operatives, and experts in the fields of counterintelligence and disinformation all fed Bennewitz fic-titious tales of dangerous extraterrestrials, of thousands of people abducted and mind-controlled in slavelike fashion by the aliens, of terrible experiments undertaken on people held below the Dulce base, and of a looming confrontation between the human race and the deadly creatures from another galaxy.

Well-placed government agents, intelligence operatives, and experts in the fields of counterintelligence and disinformation all fed Bennewitz fictitious tales of dangerous extraterrestrials [and] terrible experiments undertaken on people held below the Dulce base.

TIME TRAVEL: THE SCIENCE AND SCIENCE FICTION

That the data was all coming to him from verifiable insider sources impressed Bennewitz and led him to believe their every word—which is precisely what the government was gambling on. The government tightened the noose around Bennewitz's neck by feeding him more and more horror stories of the alien variety. And slowly, bit by bit, Bennewitz's paranoia grew. If anyone walked casually past the family home, they just had to be government agents. If the phone rang out but stopped ringing before he had a chance to answer it, then that was a sign of intimidation—from *them*. He couldn't sleep, he became stressed to the point where he required medication, and eventually he had a nervous collapse and was hospitalized. The result: he walked away from UFOs, secret projects, and cosmic conspiracies as a crushed man—which may have been the intent of the government all along.

Although the saga of Paul Bennewitz began in the latter part of the 1970s and was pretty much over by the mid-1980s, the story of the Dulce base developed legs. They are legs that walk to this very day, primarily because so many people within ufology find the tales of the underground base exciting. It really is that simple. And the government has, to a degree, continued to encourage the wilder and darker side of ufology as a means to further muddy the waters of what it is really up to in its development of new and advanced aircraft that many might perceive as UFOs. That said, though, there are those who absolutely stand by the claims that a huge, underground installation exists below Dulce. In many respects, the newer tales are even stranger and more horrifying than those that Bennewitz had shoved down his throat in the early eighties.

It's intriguing to note that Dulce's saturation in weirdness began years before Bennewitz was on the scene.

More than a decade before Bennewitz came to believe the awful rumors of Dulce were true, the U.S. government already had a stake in the area. A contingent from the Atomic Energy Commission rode into town and set up what was called Project Gasbuggy. It was a subproject of a much bigger project called Plowshare. The plan was to detonate—way below Dulce—a small nuclear device in an attempt to extract natural gas. The operation went ahead on December 10, 1967, and it worked all too well. The bomb was detonated at a depth of more than 4,000 feet. Years later, however, researchers suggested that the natural-gas scenario was a cover for something else. You may already see where this is all going. There is an enduring belief in ufology that the nuke was actually used by a panicked government to try to wipe out the alien base and the extraterrestrials said to live deep within it. Even to this day it is illegal to dig in the area on the orders of the Atomic Energy Commission, which happens to have deep ties to Area 51.

The FBI was heavily involved in the investigation of the mutilations at Dulce and has now placed its files on the mystery on its website, The Vault.

TIME TRAVEL: THE SCIENCE AND SCIENCE FICTION

Moving on, from 1975 to 1979, the town of Dulce was hit by numerous cattle mutilations. Black helicopters soared across the skies of town by night—sometimes, and incredibly, silently. Strange lights were seen flitting around Dulce's huge Archuleta Mesa. Cows were found with organs removed and blood drained from their corpses. The incisions looked as if they were the work of lasers. For those who might find all of this to be just too incredible, it's worth noting that the FBI was heavily involved in the investigation of the mutilations at Dulce and has now placed its files on the mystery on its website, *The Vault.* It's a file that reads like science fiction and runs to more than 100 pages. Notably, as we have seen, there is strong evidence that the silent black helicopters had their origins at Area 51.

In the post-Bennewitz era, other figures came forward with their own tales of Dulce and its subterranean nightmare. Whether they were telling the truth or were fed lies and disinformation by government agents is something very much open to interpretation.

One such account came from one Jason Bishop III, which is an alias for another alias, that of Tal Lavesque. No wonder the Dulce saga is so confusing. Lavesque/Bishop published what he claimed were the words of a former employee at the base, Thomas E. Castello. According to Castello: "Level 7 is worse, row after row of thousands of humans and human mixtures in cold storage. Here too are embryo storage vats of humanoids in various stages of development. I frequently encountered humans in cages, usually dazed or drugged, but sometimes they cried and begged for help. We were told they were hopelessly insane, and involved in high risk drug tests to cure insanity. We were told to never try to speak to them at all. At the beginning we believed that story. Finally in 1978 a small group of workers discovered the truth."

Then there was Branton (aka Alan B. deWalton),who also wrote about the claimed firefight that led to the hasty retreat of the U.S. military. In his controversial work *The Dulce Book,* he stated that the human body is "surrounded by the etheric 'body,' surrounded by the astral 'body,' surrounded by the mental 'body.'"

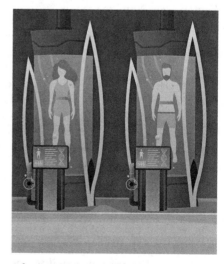

On this same issue, an insider told de Walton: "We also actually have an extra 'body,' the emotional 'body,' that the aliens don't have. This part of us constantly puts out a kind of energy they cannot generate or simulate. This emotional energy … is to them, like a potent, much sought-after drug. They can take it out of us and bottle it, so to speak…. Also during this 'harvesting,'

A former employee at a government base at Dulce claimed to have seen rows of humans there in cold storage.

TIME TRAVEL: THE SCIENCE AND SCIENCE FICTION

Greys will look directly into our eyes, as if they are drinking something or basking in light."

This sounds not at all unlike the plotline in 1985's sci-fi/horror film *Lifeforce*.

In 1991, Valdemar Valerian's book *Matrix II* hit the bookshelves. It referred to a female abductee's observations in the Dulce base, where, in "a vat full of red liquid and body parts of humans and animals … she could see Greys bobbing up and down, almost swimming."

In 2015 Joshua Cutchin penned *A Trojan Feast: The Food and Drink Offerings of Aliens, Faeries, and Sasquatch*. Cutchin's words are chilling, to say the least: "While abduction research does not overtly suggest that aliens are harvesting people for consumption, there may be a grain of truth to the report [contained in the pages of Valerian's *Matrix II*]. 'Nourishment is ingested by smearing a soupy mixture of biologicals on the epidermis. Food sources include Bovine cattle and human parts … distilled into a high protein broth.'"

Salter was sure that the so-called aliens were really biological robots created on Earth, but centuries ahead of the timeline we live in. And they come back to our time to reap us like cattle.

The fascinating thing about this chilling story is that Dan Salter agreed, to a significant degree, with just about everything. That is, except for the origins of the entities. Salter was sure that the so-called aliens were really biological robots created on Earth, but centuries ahead of the timeline we live in. And they come back to our time to reap us like cattle.

If you think this is all way too much to take in, then have a look at the official files of the FBI on cattle mutilations. They make for incredible—albeit very grisly—reading.

Cattle mutilations are a favorite topic of UFO researchers and conspiracy theorists. From the mid-1970s to the dawning of the 1980s, however, the phenomenon was also of deep interest to another body: the FBI. From January to March 1973, the state of Iowa was hit hard by cattle mutilations. Not only that, but many of the ranchers who lost animals reported seeing strange lights and black-colored helicopters in the direct vicinities of the attacks. That the FBI took keen notice of all this is demonstrated by the fact that, as the Freedom of Information Act has shown, it collected and filed numerous media reports on the cattle mutes in Iowa. The next piece of data dates from early September 1974. That's when the FBI's director, Clarence M. Kelley, was contacted by Senator Carl T. Curtis, who wished to inform the Bureau of a wave of baffling attacks on livestock in Nebraska—the state in which Curtis resided and that he represented.

At the time, the FBI declined to get involved, as Director Kelley informed the senator: "It appears that no Federal Law within the investigative jurisdiction of the FBI has been violated, inasmuch as there is no indication of interstate transportation of the maimed animals."

One year later, in August 1975, Senator Floyd K. Haskell of Colorado made his voice known to the FBI on the growing cattle mutilation controversy: "For several months my office has been receiving reports of cattle mutilations throughout Colorado and other western states. At least 130 cases in Colorado alone have been reported to local officials and the Colorado Bureau of Investigation (CBI); the CBI has verified that the incidents have occurred for the last two years in nine states. The ranchers and rural residents of Colorado are concerned and frightened by these incidents. The bizarre mutilations are frightening in themselves: in virtually all the cases, the left ear, rectum and sex organ of each animal has been cut away and the blood drained from the carcass, but with no traces of blood left on the ground and no footprints."

The senator had much more to say, too: "In Colorado's Morgan County area there has [sic] also been reports that a helicopter was used by those who mutilated the carcasses of the cattle, and several persons have reported being chased by a sim-

Many of the ranchers who lost animals reported seeing strange lights and black-colored helicopters in the direct vicinities of the attacks.

TIME TRAVEL: THE SCIENCE AND SCIENCE FICTION

ilar helicopter. Because I am gravely concerned by this situation, I am asking that the Federal Bureau of Investigation enter the case.

"Although the CBI has been investigating the incidents, and local officials also have been involved, the lack of a central unified direction has frustrated the investigation. It seems to have progressed little, except for the recognition at long last that the incidents must be taken seriously. Now it appears that ranchers are arming themselves to protect their livestock, as well as their families and themselves, because they are frustrated by the unsuccessful investigation. Clearly something must be done before someone gets hurt."

Again, the FBI—rather suspiciously, as some ranchers and media people thought—declined to get involved in the investigation of the phenomenon. It was a stance the FBI rigidly stuck to, despite collecting numerous nationwide newspaper and magazine articles on the subject, until 1978. That was when the FBI learned of an astonishing number of horse and cattle mutilations in Rio Arriba County, New Mexico. These mutilations actually dated back to 1976, and they had all been scrupulously investigated and documented by police officer Gabe Valdez of Espanola.

It was when the FBI was contacted by New Mexico Senator Harrison Schmitt (also the twelfth person to set foot on the moon, in December 1972), who implored the FBI to get involved, that action was finally taken. In March 1979, Assistant Attorney General Philip Heymann prepared a summary on the New Mexico cases for the FBI and, for good measure, photocopied all of Officer Valdez's files to the Bureau's director. Things were about to be taken to a new level.

As Valdez's voluminous records showed, from the summer of 1975 to the early fall of 1978, no less than 28 cattle mutilation incidents occurred in Rio Arriba County. One of the most bizarre events occurred in June 1976, as Valdez's files demonstrate:

"Investigations around the area revealed that a suspected aircraft of some type had landed twice, leaving three pod marks positioned in a triangular shape. The diameter of each pod was 14 inches. Emanating from the two landings were smaller triangular shaped tripods 28 inches and 4 inches in diameter. Investigation at the scene showed that these small tripods had followed the cow for approximately 600 feet. Tracks of the cow showed where she had struggled and fallen. The small tripod tracks were all around the cow. Other evidence showed that grass around the tripods, as they followed the cow, had been scorched. Also a yellow oily substance was located in two places under the small tripods. This substance was submitted to the State Police Lab. The Lab was unable to detect the content of the substance.

TIME TRAVEL: THE SCIENCE AND SCIENCE FICTION

*"Investigations around the area revealed
that a suspected aircraft of some type had landed
twice, leaving three pod marks positioned in a
triangular shape."*

"A sample of the substance was submitted to a private lab and they were unable to analyze the substance due to the fact that it disappeared or disintegrated. Skin samples were analyzed by the State Police Lab and the Medical Examiner's Office. It was reported that the skin had been cut with a sharp instrument."

Seventy-two hours later, Valdez liaised with Dr. Howard Burgess, of the New Mexico–based Sandia Laboratories, with a view to having the area checked for radiation. It was a wise move. The radiation level was double that which could normally be expected. Valdez's conclusions on this issue: "It is the opinion of this writer that radiation findings are deliberately being left at the scene to confuse investigators."

The case was not over, however. Whatever or whoever was responsible for the mutilation made a return visit. Once again, we need to take a look at the official files on the affair. In Valdez's own, official words: "There was also evidence that the tripod marks had returned and removed the left ear. Tripod marks were found over Mr. Gomez's tire tracks of his original visit. The left ear was intact when Mr. Gomez first found the cow. The cow had a 3-month-old calf which has not been located since the incident. This appears strange since a small calf normally stays around the mother even though the cow is dead."

On the matter of whether the mutilations were the work of cults or natural predators, Valdez said: "Both have been ruled out due to expertise and preciseness and the cost involved to conduct such a sophisticated and secretive operation. It should also be noted that during the spring of 1974 when a tremendous amount of cattle were lost due to heavy snowfalls, the carcasses had been eaten by predators. These carcasses did not resemble the carcasses of the mutilated cows. Investigation has narrowed down to these theories which involve (1) Experimental use of Vitamin B12 and (2) The testing of the lymph node system. During this investigation an intensive study has been made of (3) What is involved in germ warfare testing, and the possible correlation of these 3 factors (germ warfare testing, use of Vitamin B12, testing of the lymph node system)."

A further, very strange, report can be found in Valdez's files from 1978:

"This four-year-old cross Hereford and Black Angus native cow was found lying on left side with rectum, sex organs, tongue, and ears removed. Pinkish blood from [illegible] was visible, and after two days the blood still had not coagulated. Left front and left rear leg were pulled out of their sockets apparently from the weight of the cow which indicates that it was lifted and dropped back to the ground. The ground around and under the cow was soft and showed indentations where the cow had been dropped. 600 yards away from the cow were the 4-inch circular indentations similar to the ones found at the Manuel Gomez ranch on 4-24-78.

"This cow had been dead approximately [illegible] hours and was too decomposed to extract samples. This is the first in a series of mutilations in which the cows' legs are broken. Previously the animals had been lifted from the brisket with a strap. These mutilated animals all dehydrate rapidly (in one or two days)."

> *"One has to admit that whoever is responsible for the mutilations is very well organized with boundless financing and secrecy."*

As the summer of 1978 progressed, so did the number of reports where elevated radiation readings were found, as Valdez noted in his records:

"It is believed that this type of radiation is not harmful to humans, although approximately 7 people who visited the mutilation site complained of nausea and headaches. However, this writer has had no such symptoms after checking approximately 11 mutilations in the past 4 months. Identical mutilations have been taking place all over the Southwest. It is strange that no eye witnesses have come forward or that no accidents [have] occurred. One has to admit that whoever is responsible for the mutilations is very well organized with boundless financing and secrecy. Writer is presently getting equipment through the efforts of Mr. Howard Burgess, Albuquerque, N.M. to detect substances on the cattle which might mark them and be picked up by infra-red rays but not visible to the naked eye."

A lengthy document prepared by Forrest S. Putman, the FBI's special agent in charge at Albuquerque, New Mexico, was soon thereafter sent to the FBI's headquarters in Washington, D.C. It read:

"Information furnished to this office by Officer Valdez indicates that the animals are being shot with some type of paralyzing drug and the blood is being drawn from the animal after an injection of an anti-coagulant. It appears that in some instances the cattle's legs have been broken and helicopters without any identifying numbers have reportedly been seen in the vicinity of these mutilations.

"Officer Valdez theorizes that clamps are being placed on the cow's legs and they are being lifted by helicopter to some remote area where the mutilations are taking place and then the animal is returned to its original pasture. The mutilations primarily consist of removal of the tongue, the lymph gland, lower lip and the sexual organs of the animal.

TIME TRAVEL: THE SCIENCE AND SCIENCE FICTION

"Much mystery has surrounded these mutilations, but according to witnesses they give the appearance of being very professionally done with a surgical instrument, and according to Valdez, as the years progress, each surgical procedure appears to be more professional. Officer Valdez has advised that in no instance, to his knowledge, are these carcasses ever attacked by predator or scavenger animals, although there are tracks which would indicate that coyotes have been circling the carcass from a distance. Special Agent Putman then informed the Director of the outcome of Valdez's run-ins with officials.

"He also advised that he has requested Los Alamos Scientific Laboratory to conduct investigation for him but until just recently has always been advised that the mutilations were done by predatory animals. Officer Valdez

One officer theorized that the cows were being lifted by helicopter to some remote area where the mutilations took place, and then the animals were returned to their original pasture.

stated that just recently he has been told by two assistants at Los Alamos Scientific Laboratory that they were able to determine the type of tranquilizer and blood anticoagulant that have been utilized."

Putman then demonstrated to headquarters the astonishing scale of the mutilation puzzle:

"Officer Valdez stated that Colorado probably has the most mutilations occurring within their State and that over the past four years approximately 30 have occurred in New Mexico. He stated that of these 330, 15 have occurred on Indian Reservations but he did know that many mutilations have gone unreported which have occurred on the Indian reservations because the Indians, particularly in the Pueblos, are extremely superstitious and will not even allow officers in to investigate in some instances. Officer Valdez stated since the outset of these mutilations there have been an estimated 8,000 animals mutilated which would place the loss at approximately $1,000,000."

Putman additionally advised the director: "It is obvious if mutilations are to be solved there is a need for a coordinated effort so that all material available can be gathered and analyzed and further efforts synchronized. Whether the FBI should assume this role is a matter to be decided. If we are merely to investigate and direct our efforts toward the 15 mutilated cattle on the Indian reservation we, I believe, will be in the same position as the other law enforcement agencies at this time and would be seeking to achieve an almost impossible task.

"It is my belief that if we are to participate in any manner that we should do so fully, although this office and the USA's office are at a loss to determine what statute our investigative jurisdiction would be in this matter. If we are to act solely as a coordinator or in any other official capacity the sooner we can place this information in the computer bank, the better off we would be and in this regard it would be my recommendation that an expert in the computer field at the Bureau travel to Albuquerque in the very near future so that we can determine what type of information will be needed so that when the invitation for the April conference is submitted from Senator Schmitt's Office that the surrounding States will be aware of the information that is needed to place in the computer.

"It should be noted that Senator Schmitt's Office is coordinating the April conference and will submit the appropriate invitations and with the cooperation of the USA, Mr. Thompson will chair this conference. The FBI will act only as a participant."

Putman went on to describe the theories that had been advanced to try to explain the phenomenon: "Since this has not been investigated by the FBI in any manner we have no theories whatsoever as to why or what is responsible for these cattle mutilations. Officer Gabe Valdez is very adamant in his opinion that these mutilations are the work of the U.S. Government and that it is some clandestine operation either by the CIA or the Department of Energy and in all probability is connected with some type of research into biological warfare. His main reason for these beliefs is that he feels that he was given the 'run around' by Los Alamos Scientific Laboratory and they are attempting to cover up this situation. There are also theories that these are cults (religious) of some type of Indian rituals resulting in these mutilations and the wildest theory advanced is that they have some connection with unidentified flying objects."

In the closing section of his report, Putman said:

"If we are to assume an investigative posture into this area, the matter of manpower, of course, becomes a consideration and I am unable to determine at this time the amount of manpower that would be needed to give this our full attention so that a rapid conclusion could be reached. The Bureau is requested to furnish its comments and guidance on this whole situation including, if desired, the Legal Counsel's assessment of jurisdictional question. An early response would be needed, however, so that we might properly, if requested to do so, obtain the data bank information. If it appears that we are going to become involved in this matter, it is obvious that there would be a large amount of correspondence necessary and Albuquerque would suggest a code name be established of BOVMUT."

As a result of the growing concern surrounding the cattle mutilations, a conference on the subject was held April 20, 1979, at the Albuquerque Public Library.

There was a heavy concentration of FBI agents at the conference. This resulted in the preparation of the following official document summarizing the various theories, cases, and ideas advanced at the conference:

"Forrest S. Putman, Special Agent in Charge (SAC), Albuquerque Office of the FBI, explained to the conference that the Justice Department had given the FBI authority to investigate those cattle mutilations which have occurred or might occur on Indian lands. He further explained that the Albuquerque FBI would look at such mutilations in connection with mutilations occurring off Indian lands for the purpose of comparison and control, especially where the same methods of operation are noted. SAC Putman said that in order for this matter to be resolved, the facts surrounding such mutilations should be gathered and computerized.

"District Attorney Eloy Martinez, Santa Fe, New Mexico, told the conference that his judicial district had made application for a $50,000 Law Enforcement Assistance Administration (LEAA) Grant for the purpose of investigating the cattle mutilations. He explained that there is hope that with the funds from this grant, an investigative unit can be established for the sole purpose of resolving the mutilation problem. He said it is his view that such an investigative unit could serve as a headquarters for all law enforcement officials investigating the mutilations and, in particular, would serve as a repository for information developed in order that this information could be coordinated properly. He said such a unit would not only coordinate this information, but also handle submissions to a qualified lab for both evidence and photographs. Mr. Martinez said a hearing will be held on April 24, 1979, for the purpose of determining whether this grant will be approved.

"Gabe Valdez, New Mexico State Police, Dulce, New Mexico, reported he has investigated the death of 90 cattle during the past three years, as well as six horses. Officer Valdez said he is convinced that the mutilations of the animals have not been the work of predators because of the precise manner of the cuts. Officer Valdez said he had investigated mutilations of several animals which had occurred on the ranch of Manuel Gomez of Dulce, New Mexico.

"Manuel Gomez addressed the conference and explained he had lost six animals to unexplained deaths which were found in a mutilated condition within the last two years. Further, Gomez said that he and his family are experiencing fear and mental anguish because of the mutilations.

"David Perkins, Director of the Department of Research at Libre School in Farasita, Colorado, exhibited a map of the United States which contained hundreds of colored pins identifying mutilation sites. He commented that he had been making a systematic collection of data since 1975, and has never met a greater challenge. He said, 'The only thing that makes sense about the mutilations is that they make no sense at all.'

"Tom Adams of Paris, Texas, who has been independently examining mutilations for six years, said his investigation has shown that helicopters are almost always

An independent investigator, according to an FBI report, found that "helicopters are almost always observed in the area of the mutilations. He said that the helicopters do not have identifying markings and they fly at abnormal, unsafe, or illegal altitudes."

observed in the area of the mutilations. He said that the helicopters do not have identifying markings and they fly at abnormal, unsafe, or illegal altitudes.

"Dr. Peter Van Arsdale, Ph.D., Assistant Professor, Department of Anthropology, University of Denver, suggested that those investigating the cattle mutilations take a systematic approach and look at all types of evidence [and] is discounting any of the theories such as responsibility by extraterrestrial visitors or satanic cults.

"Richard Sigismund, Social Scientist, Boulder, Colorado, presented an argument which advanced the theory that the cattle mutilations are possibly related to activity of UFOs. Numerous other persons made similar type presentations expounding on their theories regarding the possibility that the mutilations are the responsibility of extraterrestrial visitors, members of Satanic cults, or some unknown government agency.

"Dr. Richard Prine, Forensic Veterinarian, Los Alamos Scientific Laboratory (LASL), Los Alamos, New Mexico, discounted the possibility that the mutilations had been done by anything but predators. He said he had examined six carcasses and in his opinion predators were responsible for the mutilation of all six.

"Dr. Claire Hibbs, a representative of the State Veterinary Diagnostic Laboratory, New Mexico State University, Las Cruces, New Mexico, said he recently came to New Mexico, but that prior to that he examined some mutilation findings in Kansas and Nebraska. Dr. Hibbs said the mutilations fell into three categories: animals killed and mutilated by predators and scavengers, animals mutilated after death by 'sharp instruments' and animals mutilated by pranksters.

"Tommy Blann, Lewisville, Texas, told the conference he has been studying UFO activities for twenty-two years and mutilations for twelve years. He explained that animal mutilations date back to the early 1800s in England and Scotland. He also pointed out that animal mutilations are not confined to cattle, but cited incidents of mutilation of horses, dogs, sheep, and rabbits. He also said the mutilations are not only nationwide, but international in scope.

"Between April and December 1975, his Bureau [the CBI] investigated 203 reports of cattle mutilations."

"Chief Raleigh Tafoya, Jicarilla Apache Tribe, and Walter Dasheno, Governor, Santa Clara Pueblo, each spoke briefly to the conference. Both spoke of the cattle which had been found mutilated on their respective Indian lands. Chief Tafoya said some of his people who have lost livestock have been threatened.

"Carl W. Whiteside, Investigator, Colorado Bureau of Investigation, told the conference that between April and December 1975, his Bureau investigated 203 reports of cattle mutilations."

One month later, the District Attorney's Office for Santa Fe, New Mexico, secured $50,000 in funding to allow a detailed study of the evidence to commence in New Mexico. Very suspiciously, when it was announced that the program was going ahead, the FBI noted that the mutilations came to a sudden halt. This gave rise to deep suspicions that the mutilators were all too human and, having heard of the planned investigation, hastily backed away until matters calmed down—which they did when the number of new reports trailed off to nothingness.

Is there really a time-travel connection to the cattle mutilations? An agenda that involves mixing animals and people in hideous ways in an effort to save the human race of centuries ahead?

With a distinct lack of new data to go on, the ambitious program was left to study a mere handful of cases—all of which it relegated to the work of predators and absolutely nothing else. Hardly surprisingly, this gave rise to the suspicions among conspiracy theorists that this was the goal all along: launch an investigation and assert that the mutations were the work of predators and nothing else, then close the investigation down. If that was the case, then it worked all too well. The world of officialdom walked away from the mutilation problem, asserting that it had resolved the entire matter in down-to-earth terms and assuring the public and the ranching community that there was nothing to worry about.

Evidently, however, there *was* something to worry about: no sooner had the project closed down than the mutilations began again. And they continue to this day. Whether the work of the government, the military, satanic cults, or deadly extraterrestrials, all that can be said of the cattle mutilations is that—officially, at least—they are no longer of any interest to the FBI or to any other arm of officialdom.

Is there really a time-travel connection to the cattle mutilations? An agenda that involves mixing animals and people in hideous ways in an effort to save the human race of centuries ahead? It would be preferable if the story was simply that— a story. But who can say?

CHAPTER 24

THE WORLD'S MOST FAMOUS TIME TRAVELER

In 2000 and 2001, the world of conspiracy theorizing was rocked when a man using the name John Titor came forward, claiming to be a time traveler from the future, specifically from 2036. Such was the fascination with Titor's story that conspiracy researchers took deep notice of what he had to say, to the point that what began as an interesting series of claims quickly became a veritable phenomenon. But was Titor all that he claimed to be? Was his story of being a member of the U.S. military true? Was he really a man from the future? Or was the whole thing a strange hoax?

There is no doubt that John Titor's claims stretch credulity to the absolute max, particularly when one considers that many of his claims did not come to pass. For his supporters, though, that's not a problem: they suspect that there may be multiple alternative timelines rather than just one—which admittedly provokes even more controversy. It all began at the dawning of the twenty-first century. It was just one year before the events of September 11, 2001, when John Titor turned up on various online forums claiming to be one of Uncle Sam's warriors—but from 36 years in the future. The story that Titor told was not a good one. In fact, it was downright grim and disturbing. Some might even call it bone-chilling.

John Titor proclaimed that a civil war would begin in
the United States in 2004 and escalate until it really
blew up in 2008. The United States, said Titor, would
fragment into five separate regions.

It's important to note that we don't even know if "John Titor" was the man's real name. Initially, the claims that he posted online went nameless. That changed when, in early 2001, Titor began posting extraordinary stories to the late Art Bell's BBS forums. The site required users to provide a name, and our alleged time traveler chose "John Titor." According to his story, Titor lived and worked in the year 2036 in Tampa, Florida. He had a mission to travel back in time to 1975, ostensibly to access certain computer-based technologies that could, in some unclear fashion, help the people of the future with their own computers.

It is hardly surprising that Titor's claims quickly became big news, particularly at Art Bell's paranormal-themed bulletin board site. When Titor's claims and predictions were made in the 2000–2001 period, they understandably shook up a lot of people. After all, he was talking about nuclear war, the collapse of much of civilization, a future that was both dangerous and dark, and a world very different from ours—in many respects, even unrecognizable. Let's see what, exactly, Titor had to say.

John Titor loudly proclaimed that a civil war would begin in the United States in 2004 and escalate until it really blew up in 2008. Such was the scale of the civil war, said Titor, that the United States would fragment into five separate regions. Things would get worse: in 2015 Russia would launch a number of nuclear missiles at the United States, causing massive death, destruction, and turmoil. Titor said that the people of 2036 referred to this brief, terrible exchange between the United States and Russia as "N Day."

He explained that the future was not set in stone:
traveling back and forth in time could result in the
creation of multiple timelines.

Of course, it's important to note that none of this ever came to pass. More significantly, Titor made no mention of 9/11—without doubt the worst attack on the United States next to the events of Pearl Harbor in December 1941. Titor, however, had a get-out clause for this highly problematic part of his overall story. He explained that the future was not set in stone: traveling back and forth in time could result in the creation of multiple timelines. In other words, Titor implied, his future and our future might be radically different. In his world, there had been a brief Third World War but no 9/11.

On this matter, Mike Suave, in his 2016 book *Who Authored the John Titor Legend?*, says: "For skeptics this represents too easy of an out, placing John's story in the loathsome category of 'unfalsifiable,' leading many to dismiss the story out of hand the moment they come across what Wikipedia lists as John's 'predictive failures.'"

Although John Titor, in 2000 and 2001, claimed to be from the future and therefore able to make accurate predictions, he failed to predict the terrorist attacks of September 11, 2001.

Titor himself acknowledged this problem with his story: "When the day comes for my 'prediction' to be realized it will either happen or not. If it does happen, then your ability to judge your environment is crippled by your acceptance of me as a 'knower of all things' and gifted with the ability to tell the future. If I am wrong, then everything I have said that might possibly have made you think about your world in a different way is suddenly discredited. I do not want either. Although I do have personal reasons for being here and speaking with you, the most I could hope for is that you recognize the possibility of time travel as a reality. You are able to change your world line for better or worse just as I am. Therefore, any 'prediction' I might make has a slight chance of being incorrect anyway and you now have the ability to act on it based on what I've said. Can you stop the war before it gets here? Sure. Will you do it? Probably not."

John Titor mysteriously vanished in 2001; his claims, however, still intrigue and entertain thousands of dedicated followers. They believe that he is not the hoaxer that many believe him to be but someone whose claims of multiple timelines in our future can explain the inconsistencies in Titor's tales. My view is that because there are inconsistencies, it means the story is bullshit.

Let's see what others who have addressed the Titor controversy think about the whole phenomenon—as a phenomenon is surely what it is.

TIME TRAVEL: THE SCIENCE AND SCIENCE FICTION

Paul Seaburn, a writer at *Mysterious Universe*, says: "'Greetings. I am a time traveler from the year 2036. I am on my way home after getting an IBM 5100 computer system from the year 1975. My 'time' machine is a stationary mass, temporal displacement unit manufactured by General Electric. The unit is powered by two top-spin dual-positive singularities that produce a standard off-set Tipler sinusoid. I will be happy to post pictures of the unit.'

"On or about November 02, 2000, that message was posted on a public internet forum by someone who first called himself Timetravel_0 and then later changed to 'John Titor' when he moved to Art Bell's BBS forum. Old geeks or computer historians will remember that the IBM 5100 Portable Computer was a 55-pound luggable with a built-in keyboard, five-inch CRT display and tape drive and was powered by a 16-bit PALM (Put All Logic in Microcode) processor. It predated the IBM personal computer by 6 years and, looking back, it's hard to imagine so little computing power in a time machine, let alone one moving not only a person but a 1967 Corvette through time. Wait, what?

"Yes, John Titor claimed his time machine or 'stationary mass, temporal displacement unit powered by two top-spin, dual positive singularities' was first installed in the rear of a 1967 Chevrolet Corvette convertible and later in a 1987 truck having four-wheel drive (is time travel that rough?). If that design sounds familiar, it's similar to the DeLorean-based time machine in *Back to the Future* which was released in 1985—before 2000 when Titor first went public but after 1975 when he allegedly picked up his 1967 Vette."

As college student Jake Kell pumped gasoline into his truck, a bald man in a business suit yelled "What year is it?" at him from across the parking lot.

TIME TRAVEL: THE SCIENCE AND SCIENCE FICTION

Surely the parallels between the fictional DeLorean of *Back to the Future* and Titor's alleged Chevrolet Corvette are not down to mere coincidence. Indeed, methinks the person behind all of this was having his or her defining moment—a moment of garbage, I should stress. On the other hand, maybe we shouldn't be too harsh. I say that for one reason: there's another strange time-travel themed saga that blends a car and other times together. We have to take a look at the work of Jason Offutt. He gives us the following:

"In 2003, Springfield, Missouri, college student Jake Kell encountered someone at a convenience store who made him wonder if time travel was possible. As Kell pumped gasoline into his truck, a bald man in a business suit yelled 'What year is it?' at him from across the parking lot. The suit, Kell noticed, wasn't typical. It was, 'along the lines of the things Teddy Roosevelt could wear.' The man yelled at him again. 'What year is it?' Kell would have panicked, except the man didn't look crazy and was obviously well dressed. He was just angry. 'Two-thousand three,' Kell told him. The man's face turned red, and he screamed the question again. Kell told him 2003 one last time, then got into his truck. When Kell looked in the rear-view mirror, the man was gone. He wasn't in the parking lot, he wasn't in front of the store, and Kell didn't see the man through the large plate glass windows on the front of the store. The man had simply disappeared."

There are also the words from writer Brent Swancer on the matter of Titor and the surrounding controversies: "Throughout these posts, Titor generated a good deal of debate as to his authenticity, and went from some random poster to somewhat of an internet sensation within a short period of time, with people calling him everything from a hoax to the real thing, and he was a recurring topic in mainstream media, on TV shows, in books, and on radio shows such as *Coast 2 Coast AM*. He was also exhaustively discussed and debated all over the internet, all a pretty impressive feat for what started out as some anonymous message board poster. John Titor would write his final post in March of 2001, in which he said some goodbyes and addressed the age old question: if time travel ever is created, then where are all of the time travelers?"

It's apt that we end this chapter with Titor himself (time traveler or master hoaxer): "In the last few days I have found your choice of topics quite interesting and from an objective viewpoint I think it collectively answers one of your own questions, 'If time travel is real, where are all the time travelers?' In the past, I have stated that quite frankly, you all scare the Hell out of me and I'm sure other temporal drivers would feel the same. But now I have an expanded explanation with two examples.

"Those two examples best define why time travelers do not show themselves. In trying to help you, we put ourselves as great risk and there's really no point to it."

"A while ago (on one of the posts), I related an experience I had with my parents while we were driving down a highway. Every now and then, we would pass someone who was in obvious distress with their vehicle. I was amazed that so many people could pass them by without stopping to help. Their explanation was fear. The risk of helping someone was too great and with today's technology, they probably had a cell phone anyway. If they didn't, the walk to a gas station would be good for them and teach them a lesson for running out of gas.

"The other example is the plight of the homeless. When you pass them as individuals on the street I see the way people selectively choose an alternate path to avoid them.

"Those two examples best define why time travelers do not show themselves. In trying to help you, we put ourselves as great risk and there's really no point to it. We know the nature of time dictates that traveling between 'exact' worldlines is impossible. Therefore, the only results we will see will be the ones we stay to see. Since worldlines, outcomes and events are infinite, we have better things to do. When I arrive in the 'new' 1998 worldline on my way home I could easily start all of this again and continue to go through the same conversations with all of the same people. However, I already know you won't pay any attention or believe me because we've already been through it on this worldline. Besides, I think the walk to the gas station will do you some good."

CHAPTER 25

TIME-TRAVELING ANIMALS

hen we think of time travel, we think of us, the human race, surfing through the centuries—maybe through millennia or even longer. But how about animals traveling through time? Of course, if such a thing has happened, then it surely must have happened via unforeseen fashions, possibly as a result of entering into a wormhole and ending up in a totally new environment—namely, ours! There is enough data on record to suggest that this particular scenario has really happened.

Imagine driving, late at night, across the foggy moors of central England and coming across what looks like nothing less than a living, breathing pterodactyl! Think it couldn't happen? It already has. From 1982 to 1983, a wave of sightings of such a creature, presumed extinct for 65 million years, occurred in an area called the Pennines. The area is known as the "backbone of England" and is comprised of rolling hills and mountains. So far as can be determined, the first encounter occurred at a place with the highly apt name of the Devil's Punchbowl on September 12, 1982. That was when a man named William Green came forward with an astonishing story of what he encountered at Shipley Glen woods. It was a large, grey creature that flew in "haphazard" style and possessed a pair of large, leathery-looking wings. The latter point is notable since it effectively rules out a significantly sized, feathery bird and does indeed place matters into a pterodactyl category.

Seventy-two hours later, a woman named Jean Schofield had the misfortune of seeing the immense beast at the West Yorkshire town of Yeadon. That the thing was

Modern-day sightings of pterodactyls in England and in the United States have given rise to the idea that the giant creatures somehow traveled through time to the present.

heading for the Leeds Bradford Airport provoked fears in Schofield's mind of a catastrophic midair collision between a passenger plane and the mighty winged thing.

Perhaps inevitably, the local media soon heard of the sightings, and the story was given pride of place in the newspapers of the day. While the theory that a large bird of prey had escaped from a menagerie or zoo satisfied the skeptics, it did not go down well with the witnesses, who were sure that what they had encountered was something straight out of the Jurassic era. Rather notably, the media attention brought forth additional witnesses, including Richard Pollock, who claimed he and his dog had been dive-bombed by the monster, which descended on the pair with alarming speed, "screaming" as it did so. Pollock hit the ground, protecting his dog beneath him. Given the fact that the creature was practically on top of him, Pollock couldn't fail to get a good look at it: he described it as reptilian, with a face that looked like a cross between a crocodile and a bat, which is actually not a bad description of a pterodactyl.

There was then somewhat of a lull in reports, but they exploded again in May 1983. There was a sighting at Thackley on May 6 by a witness whose attention to the creature was provoked by the sudden sound of heavy wings beating above. Yet again, it was a case that not only caught the media's attention but provoked others to come forward. One of them was a Mr. Harris, who said that in November 1977, at Totley, he saw just such a flying monster soaring overhead, growling as it did so. He was adamant that what he saw was a full-blown pterodactyl.

Quite naturally, further attempts were made to try to lay the matter to rest, including the amusing—but utterly unproven—theory that the pterodactyl was actually a radio-controlled model! And claims of escaped, exotic birds were once again trotted out but without a shred of evidence to support them.

TIME TRAVEL: THE SCIENCE AND SCIENCE FICTION

*If we can rule out radio-controlled models and
mistaken identity, what does that leave us with? Time
travelers, that's what we're left with.*

A few more, somewhat vague, reports trickled in. For the most part, however, the curious affair of the Pennines pterosaur was over. If we can rule out radio-controlled models and mistaken identity, what does that leave us with? Time travelers, that's what we're left with.

Joshua P. Warren is a highly respected ghost hunter from Asheville, North Carolina. Among his many investigations of the spooky kind, one in particular stands out. In the early to mid-2000s, Warren investigated and documented a huge amount of supernatural activity at the Jackson Farm in Lancaster, South Carolina. Ghostly apparitions were commonplace on the large property, particularly late at night. One of them, amazingly, was described as being a spectral, winged animal that "had a wide wingspan and a long neck with some kind of huge bird head." The terrified witness, Adam Jackson, drew a picture of the ethereal monster he encountered. It would be hard to find a better image of a pterodactyl. I stressed that Warren is a ghost hunter to show that some such cases might actually involve not time travelers but the spirits of long-dead animals. Or perhaps we're seeing a blend of both.

What about a Smilodon—or saber-toothed tiger, as it's popularly known? Let's see. About saber-toothed cats, we have these words from the University of California Museum of Paleontology at Berkeley: "Smilodon is a relatively recent sabertooth, from the Late Pleistocene. It went extinct about 10,000 years ago." According to Jenny Burrows, on a particular day in 2009, she had been walking through the woods with her pet Labrador dog, Bobbie, when the dog suddenly stopped in its tracks, whined loudly, and dropped to the floor, shaking. Thinking that her faithful pet had possibly had a seizure, Jenny quickly bent down to comfort the dog. She could then see that it was staring intently to its left. Following Bobbie's gaze, Jenny was horrified to see moving in the undergrowth what looked like a large cat—"like a mountain-lion, but it was much bigger."

Then the event got extremely strange. As the saber-toothed monster loomed fully into view and out of the confines of the bushes and undergrowth, she could see that its body seemed to be semitransparent and that "the bottom of its front paws were missing or invisible. It looked at me with a sort of surprise when it saw me watching it, and then it was gone, just like that. It was terrifying, absolutely terrifying; but it was a beautiful animal, too. Seeing it was scary, but a privilege, too."

A perfect example in the category of time-traveling animals is the account of a woman named Jill O'Brien. She had a bizarre encounter in September 2008 in Alaska's Wrangell–St. Elias National Park and Preserve. It was an encounter that I was able to personally investigate in early January 2009, as Jill had then recently moved to Oklahoma City, which is not that long a drive from my home in Arlington, Texas. It was time for a road trip and a notable interview. Jill's story was as intriguing as it was bizarre.

As Jill prepared to take one particular photo of Mount St. Elias, she heard what she described to me as "thudding" and "crunching" on the ground, coming from somewhere behind her. Her first, terrifying thought was "bear." Thankfully, it was not a bear. Jill found herself confronted by a very small mammoth, somewhere in the region of four feet in height or maybe slightly bigger.

Now, for years, reports have surfaced and circulated suggesting that, against all the odds, in certain parts of the world the mammoth survived extinction and still lives. I have investigated a lot of cases involving alleged surviving mammoths. Personally, and for the record, I don't think the mammoth is still with us. It's just not feasible. I do, however, think there is enough circumstantial evidence to suggest it may have clung on for far longer than many suspect. But that's another story for another day. Those excited by the prospect that mammoths still live today should note I'm *not* suggesting Jill saw a young living, breathing mammoth. And it's most assuredly not what Jill was suggesting, either. You'll see why.

As Jill watched in a mixture of awe and shock, the mini-mammoth raced past her and … *vanished*. Gone. When I asked Jill to explain what she meant, she said that

A hiker in Alaska's Wrangell–St. Elias National Park and Preserve encountered what appeared to be a baby mammoth racing past her, perhaps dashing briefly out of one timeline and into another.

the animal was suddenly enveloped by a small "black cloud," which "sucked into itself" and disappeared. The entire encounter lasted barely a handful of seconds at most. Jill was, however, very sure of what she had seen. In her opinion, it was nothing less than the ghost of a long-dead mammoth. Based on everything we've seen so far, though, I'm inclined to think that that baby mammoth may have briefly stepped out of one timeline and into another—and then gone back to where it originally came from.

CHAPTER 26

TOURISTS OF THE
TIME-TRAVELING TYPE

What was, and still remains, one of the most fascinating and unsettling stories of a time-traveling Woman in Black (WIB) surfaced in the latter part of 2010, but it had its origins decades earlier, in the 1920s. It was a story of truly mind-blowing and surreal proportions, and it involved swirling tales of time travel and none other than the legendary actor-director Charlie Chaplin!

The strange saga began in October 2010. That was when old and grainy footage surfaced that was connected to Chaplin's 1928 silent movie *The Circus*. Rather incredibly, the film footage in question appears to show nothing less than a black-garbed woman, holding to her ear a cell phone, and doing her utmost to ensure her identity is hidden from the camera that is filming her—something of which she is, clearly, acutely aware. Not only that, there is something profoundly unsettling about the WIB, too, as will soon become apparent. First, however, some vital background data to set the scene.

> *As we have seen time and again, where negativity and darkness lurk, so do the Women in Black.*

Chaplin's movie starred the man himself as a circus clown who can only make people laugh by accident rather than on cue. Although *The Circus* was a huge financial success—it remains the seventh-highest-grossing silent movie of all time—it was a production utterly blighted by bad luck and ill omen. As we have seen time and again, where negativity and darkness lurk, so do the Women in Black.

During the course of the production of *The Circus*, Chaplin had the Internal Revenue Service on his back. His mother, who, in a deranged state, had spent years confined to a variety of nightmarish asylums, passed away. The studio in which the movie was made caught fire, almost disastrously so, which set back filming by four weeks. The film negative was found to be scratched, but it was eventually and skillfully restored. And to cap it all, Chaplin was served divorce papers by his second wife, Lita Grey. It was after all of this mayhem and torment was over that a WIB decided to put in an appearance of the chilling kind. Her timing was appropriate for a breed of creature that seems to thrive on torment, disaster, and misfortune.

The Woman in Black in question does not appear in *The Circus* itself but in footage that was taken outside Mann's Chinese Theater in Hollywood, California, where Chaplin's movie had its premiere. It was footage added as an "extra" to a DVD release of *The Circus*. One of those who purchased and watched that extra footage was an Irish man named George Clark. As he watched the extra material, Clark saw something jaw-dropping and amazing. It was so unusual that on October 20, 2010, he uploaded the film to his YouTube page along with his own commentary, as follows:

"This short film is about a piece of footage I found behind the scenes in Charlie Chaplin's film *The Circus*. Attending the premiere at Manns Chinese Theatre in Hollywood—the scene shows a large woman dressed in black with a hat hiding most of her face, with what can only be described as a mobile phone device—talking as she walks alone. I have studied this film for over a year now—showing it to over 100 people and at a film festival, yet no one can give any explanation as to what she is doing. My only theory—as well as many others—is simple … a time traveler on a mobile phone. See for yourself and feel free to leave a comment on your own explanation or thoughts about it."

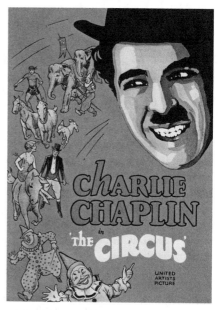

A mysterious Woman in Black, walking outside Mann's Chinese Theater in Hollywood, California, appears in footage appended to the DVD release of *The Circus*, a 1928 silent film by Charlie Chaplin.

People did exactly that. And they did much more, too: the film briefly became an internet sensation and was quickly picked up on by numerous media outlets throughout the world. At the time of this writing, the short piece of film has had close to seven million viewings on YouTube alone; never mind all the additional blogs and websites that have picked up on it. As for what the film shows, that's quite straightforward: an elderly woman, dressed entirely in black, is walking past the theater, which is displaying a prominent poster advertising Chaplin's production. Although the footage only lasts for a few seconds, it is deeply curious—and for several extremely odd reasons.

The woman's large black hat is pulled down tight around her head, to the extent that it appears she is doing her utmost to hide her face from the camera, something which she skillfully achieves. The collar of her coat—the only part of her attire that is not black—is turned up high, which further helps to obscure her face. Her coat is long and black. As for her shoes, not only are they black, they are pointed and long, to the point of almost appearing ridiculous; one might even be inclined to say they are witch- or crone-like. Now we come to the most important part of the story: the woman is clearly holding something to her left ear, using her left hand. It looks to be nothing less than a small cell phone. It, along with her hand, additionally helps to mask the woman's identity and appearance from the camera.

It almost goes without saying that down-to-earth explanations have been offered for this profoundly unusual piece of film. One of the very first things suggested was that the device in the woman's hand was a 1924 product of Siemens, something the company describes on its website as "a compact, pocket sized carbon microphone/amplifier device suitable for pocket instruments."

Siemens adds: "For a while, the carbon amplifier patented by Siemens played a major role in hearing aid technology and significantly raised the volume of hearing aids. The electrical energy controlled by the carbon microphone was not fed to the receiver directly. It first drove the diaphragm of an electromagnetic system connected to a carbon-granule chamber. Current was transmitted across this chamber from the vibrating diaphragm electrode to the fixed electrode plate. The amplified current produced mechanical vibrations in the electromagnetic hearing diaphragm that were then transmitted to the ear as sound."

Another suggestion was that the perceived "cell phone" was actually a Western Electric Company 34A carbon hearing aid, which was first manufactured in 1925. Western Electric notes that it "marketed these early hearing aids under the 'Audiphone' trade name. It was one of the few 1-piece carbon hearing aids of the time." The device measured 7.75 inches by 4 inches by 1.5 inches, and it "weighed just under 2 lbs. when fitted with batteries."

This is all well and good, and it would, to a significant degree, explain things were it not for one critical issue. A careful, close-up analysis of the footage makes it

Some suggest that the mysterious woman apparently speaking on a cell phone in 1928 is actually holding a hearing aid, such as this one, called an "Acousticon," modeled in a 1906 advertisement.

undeniable that the Woman in Black is doing something that, thus far, I have not mentioned. She is clearly *talking into the device.*

Both the Siemens and the Western Electric Company devices were designed to amplify sound for people who were hard of hearing. Neither of them had the ability—at all, in any shape or fashion—to broadcast the voice of the user to another device in the way that our cell phones do today. Plus, it's clear from watching the film that the Woman in Black is walking the street alone, so she cannot be speaking with a friend or relative next to her. The only other person in the film is a man who is clearly unconnected to her and moves out of sight before she even starts speaking. This leaves only one option: whatever the nature of the device the woman is clutching tightly to her ear, it is designed to both receive and send communications of the verbal kind over a significant distance.

There is one other anomaly worth mentioning. The size of the woman's feet, coupled with her very stocky, rotund form, gave rise to the theory that she might actually have been a man in disguise. This possibility took the affair to even weirder levels of surreal strangeness. What is even stranger is that there are more than a few reports on file of Women in Black that were actually suspected of being well-camouflaged Men in Black. Yes, the weird is about to become the downright bizarre.

Neil Arnold, a good friend of mine who has taken a keen interest in WIB reports, offers us the following:

"The *New York Times* of November 10th, 1886, speaks of 'The Woman in Black—A Queer Character That Is Causing a Sensation in Scranton (Pennsylvania).' The newspaper commented that, 'For more than a week timid and superstitious persons throughout the city have been kept in a constant state of trepidation by the appearance in various places and at unseasonable hours of an uncanny figure that is now quite generally spoken of as 'The Woman in Black.'"

Arnold continues: "The article added that, 'Two young women saw the sinister figure in the Pine Brook area after making their way home from a Saturday night 'hop.' The figure approached the women but spoke no words and gave off an air of malevolence. The young women were so horrified that they fled but the eerie woman overtook one of the women and hugged her. The terrified woman almost fainted. Her friend came to her aid but the mysterious figure in black vanished in a flash. The newspaper also goes on to mention that a few evenings previous a workman employed at the Lackawanna Iron and Coal Company had rushed to tell his colleagues that he'd seen a strange woman in black hiding in a lumber pile near the bank of the Roaring Brook.'"

The article, a copy of which was provided to me by Arnold, adds the following, which gets to the crux of the matter: "Upon investigation several men claimed to have seen a figure in black rushing towards the river from the direction of the lumber pile. A few men thought they could apprehend the woman but saw her leap into an abandoned mine. Armed with lamps the same men descended into the cave but could not find the woman. After the panic many surmised that the figure may well have been a demented local woman but *others believed the apparition to be a man dressed up*" (italics mine).

Moving on to January 1893, there was a spate of WIB activity in Rhinebeck, New York. One Gus Quirk, described as an "ex-constable" from the village, decided to look into the matter of the WIB himself, chiefly because her spectral, crone-like appearance was proving fear and sleepless nights all around the neighborhood. It's decidedly notable that the appropriately named Quirk said: "I won't stand any monkey business. I've got my suspicions. Of course they are merely suspicions and are based on what I think, but when it comes to a thing of this kind I usually think pretty nearly right thoughts. *I have thought that this woman in black was no woman at all. I had an idea that she was a boy got up to frighten people* [italics mine]. We have several boys in this village who are just about her height. I cross-questioned them pretty closely and I thought I had hit the nail on the head, but one of the villagers came in just then and shouted: 'She's been seen not more than 10 minutes ago on the river road!' Of course my suspicious persons had proved an alibi without saying a word."

Of a recent case from Blue Bell Hill, Kent, England—which involved a couple who had the distinct misfortune to cross paths with a spectral WIB—Arnold quotes what they told him: "We turned icy cold. *At first we thought it was a man dressed up* [italics mine] but now I don't know what it was. It stood on the side of the road and was beckoning us (another feature which seems common in such encounters) with a bunch of heather in its hand but I'm sure it wasn't a gypsy.'"

Note, too, the near-identical wording of "a man dressed up" in both the account given to Arnold and that reported by the *New York Times* in 1886.

One common thread was ever present in the discussion. Namely, the theory that the WIB was nothing less than a time traveler.

Whatever the true nature of the person in that priceless piece of footage—WIB or disguised MIB—when the story of Charlie Chaplin, the mysterious woman, and the attendant film footage surfaced, one common thread was ever present in the discussion: namely, the theory that the WIB was nothing less than a time traveler. Maybe that's exactly what she was. And she may not have been a solitary time surfer, either. It's time for *us*, too, to surf time: to the late 1930s.

If the 1920s-era story of the possibly time-traveling, cell phone–using Woman in Black was not enough to amaze and intrigue, there is a near-identical story from 1938 that involves a woman dressed in white. The story hit the news in 2012, two years after the Charlie Chaplin saga exploded. In this case, however, the individual causing all the mystery was a Woman in White, or WIW. The controversy began—but was initially overlooked to a significant degree—in 2012 when a brief piece of old black-and-white footage was uploaded to YouTube. It appeared to show something amazing: a young woman speaking into a cell phone! Sounds familiar? Shades of the Charlie Chaplin/WIB saga? Yes, indeed.

TIME TRAVEL: THE SCIENCE AND SCIENCE FICTION

The footage is undeniably genuine. The big question, however, is what does it show? What can be said for sure is that the WIW is holding something small and black to her ear. And she does appear to be speaking into it and engaging in conversation, all with a beaming smile on her face. As for the location, it has been identified as a Dupont factory in Leominster, Massachusetts.

It's important to note that the cell phone, as a concept and then as an initial design, did not come to the fore until the early 1970s. Its appearance was thanks to the pioneering work of Motorola. And it was not until the early 1980s that the first commercially available cell phones went on sale. They were brick-sized, cumbersome things that today look far more comical and absurd than they do revolutionary.

Among the fairly small community of people who track alleged sightings of time travelers, the debate as to what the footage showed rumbled on quietly. That all changed in 2013 when the world's media—and I do mean the *world's*—latched onto the story, the film went viral, and all sorts of theories and thoughts were trotted out to try to explain what it showed.

"Gertrude and five other women were given these wireless phones to test out for a week. Gertrude is talking to one of the scientists holding another wireless phone."

The story that was given the most publicity came via a source identified only as "Planetcheck," a YouTube user who claimed that the woman in the film was her great-grandmother, said to be named Gertrude Jones. Planetcheck said: "I asked her about this video and she remembers it quite clearly. She says Dupont had a telephone communications section in the factory. They were experimenting with wireless telephones. Gertrude and five other women were given these wireless phones to test out for a week. Gertrude is talking to one of the scientists holding another wireless phone who is off to her right as she walks by."

There are, however, problems with the claims of Planetcheck. First and foremost, she is wholly anonymous and has never come forward under her real name. Second, "Gertrude Jones" has never been formally identified. It would have been an easy thing to present evidence of Jones's existence and lay the matter to rest very quickly. Rather ironically, the media—which is usually highly dismissive when the UFO research community relies upon anonymous sources—immediately embraced the words of Planetcheck and practically cited them as established fact, which they clearly were not. Moreover, despite the words of the anonymous YouTube user, no "scientist" can be seen anywhere in the film using "another wireless phone who is off to her right."

There are other issues, too. The original posting to YouTube showing Planetcheck's words was mysteriously removed, but not before both Yahoo News and the U.K.'s *Daily Mail* newspaper copied and pasted her comments and saved them for posterity and publication, including the following: "Maybe they decided it was too far advanced for people and they abandoned the idea. Ideas are hatched, proto-types are made and sometimes like this phone they are forgotten until somebody discovers some long lost film of the world's first wireless phone and marvels at it."

Then there is the outlandish claim of Planetcheck that the family still has the device, stored in what was described as a "glass box." This is, clearly, absurd. If Du-Pont was indeed testing some sort of early cell phone way back in 1938, then the chances are that it—or they, depending on how many devices were made—would have been tested within the confines of the facility. Staff members would certainly not have been allowed to take them home, much less keep them. Such a scenario is ridiculous. It speaks volumes that Planetcheck didn't simply take a photo of the device in its box and upload it to YouTube for all to see.

In addition, there are the words of David Mikkelson, the founder of Snopes.com. He said, quite correctly: "You can take any piece of WWII footage showing someone holding something to the side of their head talking, and claim it is a time traveling cell phone user. Film clips aren't of sufficient resolution to see what the people are carrying. It could be anything from a handkerchief to a hearing aid, or who knows what. And this video is silent, so you can't even tell if the person is engaged in a two-way conversation."

Although highly skeptical of the whole affair, Mikkelson made two valid points, which echo my words above: "I doubt it would have just been handed out to a young woman working at the factory. And why isn't there documentation?" Mikkelson's question is an important one, as there is no evidence that back in 1938, DuPont was working on wireless telephones and dishing them out to its employees—even to the extent of letting them take them home!

Here's another story to share: The child of a practicing occultist and a natural medium, Felix grew up in an atmosphere that encouraged metaphysical study and introspection. After a variety of jobs in the family's hometown of Blackpool, Eng-land, at the age of 28 Felix moved to Manchester to attend university as a mature student in 1982, reading for a degree in theology and religious studies, which was awarded with upper second class honors.

After graduation, Felix volunteered as an adult literacy tutor and taught in Man-chester University's Extra-Mural Department, creating its first-ever academic course in Tarot. Felix later gained a postgraduate certificate in education (submitting a thesis on Haitian Voodoo—also a first).

Not only that, Felix had an experience that sounds very much like a time-traveling one—and with no fewer than two time travelers.

TIME TRAVEL: THE SCIENCE AND SCIENCE FICTION

During the latter part of the 1980s, Felix became chronically ill with fibromyalgia and had to cut short plans for a career in education. However, life brought other opportunities, including activism for equal opportunities as both a disabled person and as one called to live between the genders. They have a particular interest in the intersection of spirituality and gender as a cultural phenomenon and are training as a skilled helper. Felix shares life with their fiancée in Old Trafford. Felix also spends time invigilating exams, messing about in a boat, and catering to the whims of assorted cats. Not only that, Felix had an experience that sounds very much like a time-traveling one—and with no fewer than two time travelers. Their story reads as follows:

First of all, I'll set this account in context. I'm a transgender person—born female, living as androgyne and a natural psychic. My father was a practicing occultist; he was a member of the Golden dawn in the '50s and a lifelong student of magic and metaphysics. My mother attended a spiritualist church and had a number of experiences which assured her that the death of the physical body was not the end of the personality. I would describe myself as an eclectic occultist and researcher.

I was brought up to accept there will always be phenomena that we may not understand and to be critical but not dismissive of anything odd that happened to me. I was born in 1954 and raised in Blackpool, from where I moved to Manchester in 1982 as a mature student, reading Theology and Comparative Religion.

In 1981 I lived in an attic bedsit in one of the big, old houses that had been converted into flats. It was just one room, really, with a Baby Belling cooker in a former wardrobe and a sink in the corner, providing hot water through a Creda. My bed was in a corner with the window behind me and to the right. The door was past the foot of the bed on the left hand wall.

In December 1980 I had had a pregnancy terminated; this had been a stressful time, physically and mentally, but I was coping OK. I wasn't working but had worked as a bar supervisor lined up for the summer season so was taking it easy for a while.

Now, in spite of my interest in matters mysterious, UFOs and alien contact have never particularly grabbed my attention. I'd read a little about claims of alien abduction but had no books on the topic and of course there was no internet back then.

Very late one night in, I think, February 1981, I woke suddenly to see the semi-transparent form of a youth in 1930s type clothing, writing at the table under the window. He seemed unaware of me and I just thought, "That's nice; he seems happy enough. I wonder if he's a spirit or a memory imprint on the house?" [and] turned over and went back to sleep.

I don't know how long after this I was woken by the sound and "feel" of visitors. I sat up in bed and saw two men in the room, absolutely solid and

"Very late one night in, I think, February 1981, I woke suddenly to see the semi-transparent form of a youth in 1930s type clothing, writing at the table under the window."

"real," assumed they were police and wondered how the hell they had got in. I recall saying, "What?" and they approached the bed. I lay down, grabbing the quilt in fright, but they were absolutely impassive and did not address me directly. One took a gadget a bit like a mobile phone in appearance and "scanned" me with it. He told his companion—I'm not sure whether I heard this with my ears or telepathically—that this was a female 20–30 years and was infertile due to contraception (or a term meaning the same). I had recently started using the Pill.

The other man took a stroll around the room, examining its contents. He seemed very amused by anything that plugged into the wall—the Creda heater and my radio-cassette (the latter is especially important)—and commented, smiling, "I can't believe they're still harnessing it from an external source." I felt that he was familiar with people generating their own electricity via their bodies.

All the time this was going on I couldn't move or speak—it was as if they held me in a trance or catatonic state.

The next thing I knew it was morning and I was alone. I reached down, shaken, to switch the radio on and was surprised to hear just a load of static. On inspection I saw it had been tuned to an area where there is NO station. Now, I always had it tuned to Radio 1 (a lot of young folk did then) and never touched it. It was the last thing I recall the man picking up.

It wasn't until later that evening that I put together the pieces of the visit and thought, "Men in Black!"

In appearance they were very similar and dressed identically, like twins. Both around 5'8–5'9 in height and slim build. Their complexion was a bit olive/jaundiced, eyes I'm not sure and what hair I could see was black. They wore beige, straight cut raincoats over dark trousers and shoes with white shirts and black ties and each had a beigey-brown fedora. They looked odd and retro and had an aura not of menace but of indifference to humans; I felt they looked at us as specimens in a laboratory.

I have never told anyone but my parents this story and always had the feeling I was not meant to discuss it, at least for a number of years. As decades have passed, I thought I would share it with you. I hope you found it interesting!

Again, many thanks for your excellent book on this topic—I'm presently torn between the tulpa and the time-traveler models but I'd rather they didn't come back to tell me. ;-)

Best wishes, Felix

CHAPTER 27

THE LOCH NESS DOORWAYS

When we think of Loch Ness, Scotland, it's all but certain that the first thing that will spring to mind is the Loch Ness Monster. Or, to be correct, the Loch Ness Monsters. A large number of people claim to have seen the long-necked, humped leviathan of the deep. Some have even photographed and filmed it. Nessie, as the beast is affectionately known, has been a staple part of the world of the unexplained since 1933, when the phenomenon of the monster exploded in spectacular, planetwide fashion. Since then, millions of people have flocked to the shores of the 22.5-mile-long and 744-foot-deep loch with the hope of seeing the elusive creature. Attempts have been made to seek out Nessie with sonar equipment, aircraft, balloons, and even submarines.

Theories abound as to what Nessie is—or, far more likely, what the Nessies are. Certainly, the most captivating theory, and the one that the Scottish Tourist Board, moviemakers, and the general public find most appealing, is that the monsters are surviving pockets of plesiosaurs. They were marine reptiles that the domain of zoology assures us became extinct tens of millions of years ago. The possibility that the monsters are actually giant-sized salamanders holds sway in more than a few quarters, as does the idea that perhaps massive eels are the culprits. Then, there are scenarios involving sturgeon, oversize turtles, catfish, and even crocodiles, giant frogs, and hippopotami!

Numerous Nessie enthusiasts, investigators, and authors have spent years—decades, in some cases—pursuing their quarry. They have done so in a fashion that uncannily mirrors the obsessive actions of the fictional Captain Ahab in Herman

Suggestions for the nature of the Loch Ness Monster—or Monsters—include scenarios involving plesiosaurs, giant salamanders, sturgeon, oversize turtles, catfish, crocodiles, giant frogs, and hippopotami—and a time portal at the loch.

Melville's acclaimed 1851 novel *Moby Dick; or the Whale*. But it's all, and *always*, to no avail. No matter the number of days, hours, weeks, and years spent, and no matter just how advanced the technology utilized to find the animals might be, it forever ends in failure. After more than 80 years of intensive investigation, the Nessies still evade capture, discovery, or classification. Is this all down to sheer bad luck and inept investigations? Certainly not. Rather, it's a case of people looking for the answers in the completely wrong direction.

Over the decades numerous books have been written on Nessie. For the most part, they all follow the same path. It's a path that is becoming more and more weary, predictable, repetitive, and worn as the years go by. Those books typically take a near-identical approach: they chronicle the most famous sightings, the chief players in the saga, and the theories that exist to explain the monster—and then they leave it at that. All of which brings me to the theme of the book you are now reading in relation to the Nessies—that is, the connection between the loch, the creatures, and time travel. We begin with a tale of time travel at the loch from an acclaimed writer.

One of the strangest stories of paranormal weirdness at Loch Ness comes from an author on all manner of mysteries, Andrew Collins. As Christmas 1979 loomed, Collins spent a week in Scotland with his colleagues Graham Phillips and

Martin Keatman, investigating the Nessie enigma. It involved interviewing witnesses, spending time poring over old archives in Inverness's library, and checking out the loch itself. It was while they were deep in the heart of their investigation that the trio uncovered a very weird story. Back in the early eighteenth century, a young couple inexplicably vanished while riding a horse and trap near Loch End, on the south shores of Loch Ness. Rumors circulated that the two were either murdered or abducted. Neither the horse nor the trap were ever seen again. It would have remained a complete mystery were it not for one thing—a very uncanny thing.

Several of the locals recalled old tales of the events of a century earlier, including the missing pair of young lovers. Was this, perhaps, a case of a slip in time having occurred?

More than 100 years later, at the height of a tumultuous thunderstorm, a young man and woman walked into a local almshouse, inquiring if the priest who oversaw it would give them shelter for the night, which he did. The priest couldn't fail to see that the pair was dressed in the kind of clothing that was popular around a century or so earlier. Plus, they seemed very confused, dazed, bewildered, and completely unable to explain where they were from. They remained in that odd, altered state for a couple of days, after which they simply walked out of the almshouse and were never seen again. When the story got out, however, several of the locals recalled old tales of the events of a century earlier, including the missing pair of young lovers. Was this, perhaps, a case of a slip in time having occurred? It might have been.

Moving on: During the course of his investigations into the world of the U.S. government's remote-viewing programs, the late conspiracy theorizer and author Jim Marrs learned something incredible. The U.S. government, in the 1980s, had a secret team that had focused their psychic skills on the creatures. In doing so, they stumbled onto something amazing, which added much weight to the argument that the Nessies were not what they seemed to be. Marrs said that over the course of a number of attempts to remote view the Nessies, the team found evidence of what appeared to be physical, living creatures—ones that left wakes and could be photographed and tracked. They even prepared drawings that suggested the Nessies might be plesiosaurs. There was something else, too: the ability of the creatures to vanish—as in dematerialize.

The remote viewers were in a collective quandary: their work certainly supported the theory that some seriously strange creatures lurked deep in Loch Ness, but they were creatures that seemed to have supernatural and abnormal qualities about them. Of course, in light of the earlier story of time traveling at Loch Ness,

TIME TRAVEL: THE SCIENCE AND SCIENCE FICTION

you could make a case that the reason why the Nessies are so elusive is not because they are paranormal in nature but because they have the ability to surf time—just like that eighteenth-century couple referenced above—and exit the present for the future. And return again.

I could also offer yet another reason in support of a time portal existing at Loch Ness. There is the story of a man named Peter Smithson, who told ghost investigator Bruce Barrymore Halpenny of his encounter with a ghostly airman at the loch in 1978. Smithson said it was early one morning, just as dawn was breaking, when he saw someone coming toward him from the depths of Loch Ness. Smithson's first reaction—quite a natural one—was to assume there had been some kind of accident. It was easy to understand why Smithson assumed that, as the man before him was dressed in military clothing and was dragging behind him a parachute. But what baffled Smithson was the fact that the uniform the man was wearing was clearly out of date. It was far from modern-looking and more befitted the era of the 1940s, when the world was engaged in trying to defeat the hordes of Adolf Hitler. Smithson shouted to the man to see if he was okay. The response Smithson got was an eerie one: the man slightly turned and pointed toward the waters of Loch Ness. Smithson said that the man suddenly dematerialized, leaving him with a "funny feeling" and a suspicion that what he had seen was "a ghost airman." He commented: "What a damn fool I felt, confronted by a ghost, my camera around my neck, yet I never had an inkling to take a photo."

Several encounters of this type at Loch Ness have led a number of researchers to suggest there is a time loop at the loch, which endlessly plays over and over again.

Several encounters of this type at Loch Ness have led a number of researchers to suggest there is a time loop at the loch, which endlessly plays over and over again. Now I'll tell you the backstory of that "forever rotating pilot," as I call it.

It was in 1976 that the remains of a Second World War–vintage aircraft were found in Loch Ness. It was a Vickers Wellington bomber, the brainchild of one Barnes Wallis, a man who created a revolutionary "bouncing bomb" during the hostilities with the Nazis. The aircraft was designed to destroy German dams. The Dam Busters, as the team that dropped the bombs was known, succeeded beyond their wildest dreams, leaving a significant scar on Hitler's war machine. The British Royal Navy conclusively identified the aircraft as a Wellington. They even managed to identify its serial number: N2980. There was nothing particularly mysterious about N2980: as far back as late 1939, it had taken part in no less than 14 bombing missions over Germany before being transferred to Lossiemouth, where it was then used to

TIME TRAVEL: THE SCIENCE AND SCIENCE FICTION

train newly drafted aircrews. Official records on the crash were quickly accessed and told a memorable story.

Military records showed that the aircraft ditched into Loch Ness on December 31, 1940—New Year's Eve—after experiencing problems with one of the engines during a turbulent snowstorm, high above the domain of monsters. It was when the crew was over the Monadhliath Mountains, to the southeast of Loch Ness, that problems began. The problems led squadron leader Nigel Marwood-Elton to give a hasty order to jump ship, so to speak: four crew members quickly parachuted out of the plane. Tragically, one of them, the 20-year-old rear gunner, Sergeant John Stanley Fensome, was killed when his parachute catastrophically wrapped itself around one of the wings of the doomed plane.

While the crew was racing to exit the aircraft, Marwood-Elton and the copilot, named Slater, stayed on board, struggling to control the aircraft as the dusk skies threatened to give way to darkness. With the snow hammering down and a powerful wind blowing, they maneuvered the plane closer and closer to the loch and, incredibly, actually managed to land it on the surface of the water, ditching near Urquhart Castle. With water already flooding into the aircraft from all corners, they scrambled for an onboard dinghy. The two then clambered out of the plane and onto the starboard wing, where they blew up the dinghy and used it to row to shore as the aircraft was swallowed up by the waves, practically intact. As they reached land, the two men managed to flag down an astonished truck driver, who quickly drove them to Inverness—no doubt for a couple of wee and hearty drams to help steady their nerves.

It wasn't until September 1985, amid more than a few hazards and hiccups, that the bulk of the aircraft was finally raised from the water, with the recovery of

Paranormal events at Loch Ness include the loss of a Vickers Wellington bomber in the loch on December 31, 1940, resulting in the death of one of the crew—and his apparent reappearance decades later.

TIME TRAVEL: THE SCIENCE AND SCIENCE FICTION

numerous scattered fragments continuing into 1986. Incredibly, and almost unbe-
lievably, the taillights of the plane were still in good, working order. The well-pre-
served remains of N2980 can now be seen on display at the Brooklands Museum,
Surrey, England.

Now we come to an affair that ties together the time-traveling Men in Black
and Loch Ness. On the night of June 2, 1973, Ted Holiday, a renowned Nessie-
seeker—did nothing less than try to exorcise the beasts of Loch Ness and banish
them to the supernatural realm he believed they came from. It was a tumultuous
night, one which saw him take to the dark waters and try to forever eject the monsters
from the legendary loch. By the end of the night, he was emotionally exhausted,
fraught, and frazzled. The day after, however, proved to be even more terrifying.

It was while he took a walk down to the loch's shoreline shortly after breakfast
that Holiday encountered something terrifying. At first, everything seemed normal:
the skies were clear, the sun was shining, and the oppressive atmosphere of the night
before was well and truly gone. As Holiday looked around, however, he could see—
at a distance of perhaps 30 meters—a man dressed completely in black. The man
was staring right at him from the top of a small peak that overlooked the lair of the
long-necked monsters. On seeing the man, said Holiday, highly astutely, "I felt a
strong sensation of malevolence, cold and passionless."

> *On seeing the man, said Holiday, highly astutely,*
> *"I felt a strong sensation of malevolence,*
> *cold and passionless."*

That was barely the start of things: despite the sense of dread that had enveloped him, Holiday decided to walk toward the man and confront him. What a most
unwise choice that was. The closer Holiday got, the weirder things became. Holiday
could see that the stranger in black appeared to be dressed, neck to foot, in some
kind of tight-fitting, shiny black plastic. A black helmet of sorts, vaguely like the
kind of thing a biker would wear, covered his head. A thick pair of tinted goggles
was all that Holiday could see of the man's face; a black cloth was wrapped around
his nose, face, and chin. Even his hands were covered.

Holiday sucked in a deep breath and took more steps forward, his heart thump-
ing and adrenaline pumping. Holiday called out to the man but got no reply: the
MIB stood his ground, as if goading Holiday to come closer, which is exactly what
he did. Holiday was plunged into a state of cold fear when, as the two were within
10 feet of each other, he could now see that there appeared to be *no eyes behind the
goggles*—just an impenetrable nothingness. Suddenly, the air was filled with a shrill

whistle, and the MIB vanished into oblivion before a stunned Holiday. Now almost overwhelmed by fear, Holiday half stumbled and half ran for safety. He did so for around a couple of hundred feet before stopping and looking back. The MIB was still not to be seen. Holiday sighed with relief. He should have realized, however, that the Man in Black was not yet done with him.

Exactly one year later, at precisely the same spot where he had encountered the MIB the previous year, Holiday suffered a serious heart attack. He perceived this as a warning to keep away from Loch Ness and its paranormal denizens. If only he had done as he was told. Holiday, however, was not going to let a heart attack get the better of him and his research. Once he had recovered a few months later, Holiday was back to his Nessie studies. It was, however, all too late. In 1979 Holiday had another heart attack. That one killed him. We have to wonder: did the Man in Black know that Holiday would die exactly one year later, and on the exact same site? Of course, if he was a time traveler, he would know the exact time and date of Holiday's medical emergency.

CHAPTER 28

BLACK-EYED CHILDREN AND SHADOW PEOPLE

There can be no doubt that one of the creepiest phenomena to have surfaced in recent decades is that of what have become infamously known as the Black-Eyed Children (BEC). It would be overly simplistic to suggest they are merely the offspring of the Men in Black and the Women in Black. Admittedly, though, there are deep similarities, as will soon become apparent. Like the MIB and the WIB, the BEC are definitive drainers of energy. Before we get to this aspect of the story, let's first take a look at how, and under what specific circumstances, the Black-Eyed Children came to prominence. We'll also take a look at a few classic cases. None of them are particularly heartwarming (as if you really needed to be told that).

Although today there are people who claim to have seen the BEC in the period from the 1930s to the present day, the first reported case didn't surface until January 1998. The location was the Lone Star State, specifically the city of Abilene. The story revolves around a man named Brian Bethel, a journalist who works for the *Abilene Reporter-News*. It was late one night when Bethel's life was changed and he came to realize that there are dangerous, supernatural entities in our midst.

*Something about these kids rang alarm bells in
Bethel's head.*

It was close to 10:00 p.m. when Bethel had the kind of close encounter that one never, ever forgets. He had pulled up at a mall not too far from his home to deposit a check in a mailbox. All was quiet and dark. Bethel, using the lights of the mall to illuminate the interior of his vehicle, was writing the check when he was rudely interrupted. He jumped with surprise at the sight of a couple of kids who were standing next to the car on the driver's side. Something about these kids rang alarm bells in Bethel's head. In fact, as Bethel would imminently learn, things were wrong in the extreme.

Bethel stared at the pair and couldn't fail to see just how incredibly pale the face of one of the boys was. The other had what Bethel described as olive-colored skin. Both boys were around 10 to 14 years of age, Bethel estimated, and both were dressed in pullovers. Only one of the two boys spoke; he claimed that they wanted to see a movie, *Mortal Kombat*, at the local cinema. But there was a problem: they had left their money at home. Could Bethel take them to their homes to get some cash? Bethel instantly realized that this whole situation had an air of dark and disturbing theater about it. There was an undeniably unsettling agenda at work, and it had absolutely zero to do with watching movies.

> **Bethel found himself almost mind-manipulated, to the point where, to his horror, he could see that his hand was heading to the driver's-side door with the intent of opening it but without his control.**

Bethel awkwardly hemmed and hawed for a few moments, which caused the talkative boy to become ever more insistent that Bethel let them in the car. Then things became downright eerie and chilling: Bethel found himself almost mind-manipulated, to the point where, to his horror, he could see that his hand was heading to the driver's-side door *with the intent of opening it but without his control*. Bethel, fortunately, broke the enchanting spell and didn't open the door after all. This clearly incensed the talkative boy, who amped up the pleas to allow them in. It was then, for the first time, that Bethel finally got a good look at their eyes. They were completely black.

The two boys realized that by now, they were losing the battle to be permitted into Bethel's car. On this point, Bethel said that the boy who did all of the talking "wore a mask of anger." The boy, now displaying a look on his face that was part frustration and part anger, almost shouted: "We can't come in unless you tell us it's okay. Let us in!" Bethel, terrified, did nothing of the sort. What he *did* do was slam the car into reverse and head for home, completely panicked by the whole thing. Oddly, as he drove away, Bethel looked back, only to see that the boys were nowhere in sight. In what was clearly and impossibly quick time, they had vanished. Thus was quickly born a legend. It's a legend that shows no signs of stopping. It's only getting worse.

Tales of what have become known as the Black-Eyed Children have now been circulating for a few decades. The witnesses are many. The cases are creepy and bizarre. But it's fair to say that we still don't have a solid answer as to what they are or what they want. It's worth noting an intriguing UFO connection to the very strange saga of the Black-Eyed Children: they have deep parallels to the notorious time-traveling Men in Black, who have, for decades, terrorized and intimidated into silence numerous witnesses to UFOs. Just like the BECs, the Men in Black have a propensity for knocking on doors late at night. The two mirror each other in other fashions, too: like their child counterparts, the MIB will not enter a home until they are specifically invited to do so (suggesting shades of vampire lore). Both the MIB and the BEC wear black. Both wear headgear: hoodies for the Black-Eyed Children and 1950s-era fedoras for the MIB. And there is another intriguing issue: very often the Men in Black wear thick, wraparound, black sunglasses.

I have to wonder if the sunglasses are worn not just to provoke a menacing atmosphere but to mask a pair of emotion-free, solid black eyes of the type that the Black-Eyed Children are described as having. In that sense, perhaps the Men in Black and the Black-Eyed Children are one and the same: one in adult form and the other in child form, but both following the same secret agenda—whatever that may really be. So, where does that leave us? Well, it admittedly leaves us with a lot of questions and not a great deal else. But they are questions that a certain acclaimed and dedicated researcher has done his very best to resolve.

Undoubtedly the leading investigator in this field, and the author of the book *Black Eyed Children*, David Weatherly has collected and studied dozens of such cases, demonstrating the sheer scale of these weird and unsettling events. David's work

Both the MIB and the BEC wear black. Both wear headgear: hoodies for the Black-Eyed Children and 1950s-era fedoras for the Men in Black.

TIME TRAVEL: THE SCIENCE AND SCIENCE FICTION

has addressed a variety of theories for the origins and presence of the BECs: that they might be demonic in nature, predatory spirits, or possibly even products of the human mind, given some semblance of physical life in the real world. Notably, David has also spent much time researching the theory that the Black-Eyed Children have nothing less than an alien lineage. David cites the case of two women, Ann and Marcia, both of whom have had alien abduction experiences and both of whom have also encountered the BECs—and, in Marcia's case, in the now-typical "knocking on the door and demanding to be let in" situation.

Of strange cases such as these, David says: "Why would an alien race want, or need, to create hybrid beings? There are a number of proposed theories to explain why such an experiment may be taking place. These theories range from a dying race that needs an injection of fresh DNA in order to survive, to time travelers attempting to correct a mistake made in their past (which is our present)."

These further words from David fit comfortably with what we have seen so far: "An even more sinister theory that has followers, postulates that the greys are actually attempting to take over our world by slowly breeding us out and phasing in beings with alien DNA. Over time the human race will fade away to be replaced by the 'superior' DNA of the grey aliens, adherents say."

A reader named Ron told me that a week or so after reading my *Paranormal Parasites* book (it was published in May 2018, and Ron bought his copy a couple of weeks later), he woke up in the dead of night with a pounding headache, a dry mouth, and a rash on his arms (which was gone by morning). Worse still, he was unable to move. This is a real phenomenon that many of you will have heard of: sleep paralysis. As is typical in such cases, Ron immediately found himself filled with fear. This is hardly surprising: standing at the side of the bed were two young children, maybe 10 or 11 years in age, who stared at Ron through completely black eyes. Their skin was pale, and they were both dressed in black hoodies. In seconds, Ron's bedroom was flooded with apocalyptic images of a flattened city, of millions dead, and of the sky filled with blackness. Ron knew what the cause was: a huge, nuclear mushroom cloud dominated what was left of the city. Suddenly, the image was gone—as were the BEC. Ron told me that he had near-sleepless nights for the next three nights. Thankfully, the BEC did not return. Nor did the nightmare.

Adrian Clarke of England told me a similar story. In Adrian's case, it was around three o'clock one morning in September 2018, and he too suddenly found himself wide awake and with a pair of Black-Eyed Children looking at him from the foot of the bed. Both had the completely black eyes but were dressed in orange robes. Suddenly, and similar to Ron's experience, Adrian briefly found himself deep in the heart of a nightmarish world of billions dead and Earth devastated. He fought

TIME TRAVEL: THE SCIENCE AND SCIENCE FICTION

against the paralysis and finally managed to break the spell. In seconds, the image was gone—as were the BEC.

Moving on, there is the account of Jennifer, whom I met at the September 2018 Mothman Festival in Point Pleasant, West Virginia, and who lives in Point Pleasant. After I gave my lecture on my book *The Black Diary*, Jennifer came up to my table and shared with me her experience, which occurred in April 2018. She too was woken from her sleep by a pair of pale kids with huge, dark eyes. One of them pointed to one of the walls in Jennifer's bedroom, which was suddenly transformed into an approximately eight-foot-by-eight-foot image of a nuclear explosion. Jennifer told me that, for reasons she wasn't sure of, she somehow knew that the city was Milwaukee, Wisconsin. She has never been to Milwaukee, nor does she have friends or family there. Like Ron and Adrian, Jennifer was shaken to the core by her experience.

I have several more cases in which witnesses to the BEC—after their encounters with those dangerous kids—had extremely vivid dreams about future events. Five of them were of vivid, worldwide, nuclear war in the twenty-first century—a war that all but destroys the Western Hemisphere and ravages the rest of the planet. Interestingly, and disturbingly, in all five of the cases, North Korea turned out to be the culprit that set the deadly ball rolling, with not a single chance of stopping it. Billions were killed. The atmosphere was black for decades—a nightmarish phenomenon known as a "nuclear winter," the period that, theoretically, will occur if we have a worldwide nuclear war and millions of tons of radioactive dust is thrown into the atmosphere.

Are the Black-Eyed Children warning us of a future that is destined to destroy us?

In some way, these nightmares of disastrous future times mirror what we learned about the multiple stories of dreams of a radioactive apocalypse in 2017. Are the Black-Eyed Children warning us of a future that is destined to destroy us? Or, keeping the matter of timelines in mind, are the BEC trying to tell us—in their own eerie, strange way—that this is only one possible scenario—a scenario that can be avoided if we do our utmost to avoid the worst? Let us all hope for the latter.

Possibly connected to the Black-Eyed Children are the Shadow People. As with the BEC, those who encounter the Shadow People have terrifying dreams and visions of future events, all of a catastrophic, apocalyptic nature, suggesting that the

Shadow People might be travelers from a future time. In the last decade or there-abouts, a great deal of research has been undertaken into a phenomenon that is known as the Shadow People. Their title is most apt, as they appear in the form of a shadow, usually in peoples' bedrooms while they sleep. Many of these entities, but certainly not all of them, have one thing in common: they wear an old-style fedora hat and a black suit. In that sense, there are deep similarities between the Shadow People and the Men in Black of UFO lore. It's fair to say, though, that for the most part, that's where the similarities end. Apart from one important, extra thing: like the MIB, they are intent on terrorizing those who they maliciously target. And, they do an extremely good job of it, too, unfortunately for us.

Precisely who or what the Shadow People are is a matter of deep debate. At *Psychology Today*, Katherine Ramsland, Ph.D., has an interview with the late, and much missed, occult writer Rosemary Ellen Guiley. In that interview, Rosemary says: "I discovered that many Shadow People experiencers are also ET experiencers, especially abductees. Through a long process, I concluded that Shadow People are a shape-shifted form taken by Djinn. Therefore, there is a profound connection between Djinn and bad hauntings and ET abductions. Furthermore, the footprints of the Djinn are evident throughout our mythologies about ancient aliens and gods. The picture that emerged is of a major Djinn involvement in all of our entity contact experiences throughout history."

Stephen Wagner at *LiveAbout*, in an article dated January 6, 2019, says of this creepy phenomenon: "Perhaps this is an old phenomenon with a new name that is now being discussed more openly, in part thanks to the Internet. Or maybe it's a phenomenon that, for some reason, is manifesting with greater frequency and intensity now. Those who are experiencing and studying the shadow people phenomenon say that these entities almost always used to be seen out of the corner of the eye and very briefly. But more and more, people are beginning to see them straight on and for longer periods of time. Some experiencers testify that they have even seen eyes, usually red, on these shadow beings."

Jason Offutt is an expert on the subject and the author of a 2009 book titled *Darkness Walks: The Shadow People among Us*. He says there are eight different kinds of Shadow People—at least,

As with the BEC, those who encounter Shadow People have terrifying dreams and visions of catastrophic future events, suggesting that the Shadow People might be travelers from a future time.

they are the ones we know about. He labels them as benign shadows, shadows of terror, red-eyed shadows, noisy shadows, angry hooded shadows, shadows that attack, shadow cats, and the Hat Man. Certainly, the latter category—that of the Hat Man—is the one which is most often reported from all across the planet. Imagine a silhouetted character that had stepped out of a 1940s- or 1950s-era, black-and-white film noir, and you'll have a good idea of what the Hat Man looks like. That he is entirely shadowlike in nature only adds to the menace.

He says there are eight different kinds of Shadow People that we know about: benign shadows, shadows of terror, red-eyed shadows, noisy shadows, angry hooded shadows, shadows that attack, shadow cats, and the Hat Man.

Heidi Hollis is an expert on this topic, too, having penned a 2014 book on the subject titled *The Hat Man: The True Story of Evil Encounters.* She has collected literally hundreds of reports of encounters with this particularly dangerous shadow-thing. One example from Hollis's files that appears in her book reads as follows: "Dear Heidi, I was maybe 5 years old when The Hat Man started to visit me. Every night I would lay in the top bed of my bunk bed and watch as my door would crack open for him to creep inside. As high up as I was, I would still have to look up to see him and I would freeze in horror at the sight of him." According to the witness, the Hat Man would lunge toward her at a phenomenal speed and always warn her, "One day I will have you!" To be sure, this was a deeply traumatic series of encounters for a young girl, who never forgot the torment and the terror. You have been warned!

CHAPTER 29

BIGFOOT, WORMHOLES, AND TIME PORTALS

The Bigfoot creatures are undoubtedly the most well-known cryptids—animals not proven to exist—in the field of monster hunting. I would place the creatures of Loch Ness, Scotland, in second place and the Mothman third. Having taken part in expeditions in search of Bigfoot, written a book on the subject, been on various TV shows discussing it, and lectured on the topic, I think I can say that I know a fair amount about the subject. One issue, however, remains to be addressed: Why is that we can't find hard proof that the creatures are real? The question is not a simple one. In fact, it takes us down some very strange avenues. Reports of Bigfoot date back decades. We can push things even deeper into the past by taking a look at the many and varied nineteenth-century newspaper articles on so-called wild men encounters. Huge numbers of people have seen the Bigfoot creatures. Some of the luckier creature seekers have taken photos and shot film footage of the animals, which is usually subjected to a wealth of debate and controversy. Even audio recordings of the animals have been made. Some of that material, in my view, is pretty impressive. There's a big problem, however. It's Bigfoot's incredible elusiveness, which is something that prevents us from getting the proof we seek.

Here's the reality of the situation: Bigfoot creatures have been seen throughout almost all of the United States. They are massive animals, around six and a half to seven feet tall. In some cases, they're described as being even taller than that. In other words, they shouldn't be hard to see. They tower above us. Yet they seem to have a

Bigfoot creatures seem to have a never-ending "here one second and gone the next" aspect to them—perhaps evidence of dimension-hopping abilities.

never-ending "here one second and gone the next" aspect to them. Now we come to the matter of proof—or the lack of it.

While I absolutely conclude that the creatures do exist, I also conclude that Bigfoot is more than just an unknown North American ape. Despite all of the data, testimony, photos, and footage, we lack (to a 100 percent degree) any hard proof for the existence of the animals. Despite what has been loudly claimed at times, there is *no* Bigfoot DNA to be studied. None. Period. No teeth or nails to make a case. No hair or fur that can be said to be truly anomalous. It's all about "maybe," "possibly," and so on. On top of that, there's the matter of the creatures getting shot. Yep, there are such claims. Show me the body? It hasn't been done, despite all of the fieldwork and expeditions. Not even once. Also on the matter of shooting the beasts, it's a baffling fact that bullets seem to have no effect on the Bigfoot creatures.

Have you ever heard of a Bigfoot being hit by a vehicle? Yes, there are some such reports and testimony. Here's the problem, though: so far as we know, there is no absolute proof to show that a Bigfoot has ever been struck and killed by the driver of a vehicle. Most of us have never seen a grizzly bear in the woods. The same is the case when it comes to mountain lions. Nonetheless, some people *are* lucky enough to see grizzly bears and mountain lions, and now and again, they *do* get hit and killed by drivers on the roads. The evidence exists. Bigfoot, however, eludes getting killed on the roads 100 percent of the time. I can accept that Bigfoots carefully avoid getting killed—*most* of the time—along the winding roads of dense forests. But with a perfect success rate on *every single occasion*? For me, that's stretching things way too far. Why are the creatures almost always solitary? Why have we never come across a baby Bigfoot that can be shown as proof? Or a juvenile? Certainly, there are such accounts out there. But proof on show? Nope. None. How can such massive beasts, which are seen all across the United States, *never, ever* be caught or killed by us? Or, when such a claim is made, why can the source never provide some sort of hard evidence?

These points bring us to the weirder side of the Bigfoot phenomenon. Like it or not (and most Bigfoot seekers don't like it, which is just too bad), there are reports of Bigfoot vanishing in a flash of light. Other witnesses have said the animals disappeared, literally, right in front of them. Some say the animals can temporarily disable us by hitting us with infrasound, allowing them to make their escape. Now, as I see it, there is no good reason as to why people should be seeing Bigfoots that can dematerialize or that appear to step into another realm of existence. We're talking about the likes of dimension-hopping animals. Nonetheless, people do report such things, even though the numbers are admittedly small.

Why is there no undeniable proof for the existence of Bigfoot? The answer: because Bigfoot is not what it appears to be.

Why is there no undeniable proof for the existence of Bigfoot? The answer: because Bigfoot is not what it appears to be. Pursuing the Bigfoot mystery from a zoological approach is ridiculous. Bigfoot is too elusive and too weird. Of course, there's no proof, either, for those stories of bullets having no effect on Bigfoot or of the monsters disappearing before the eyes of shocked people. The fact is, though, that if Bigfoot is only an ape, then there shouldn't be *any* reports *at all* of dematerializing Bigfoot. Nor should there be any claims of Bigfoots seeming to make their escapes via portals while strange lights hover in the sky. This all takes me to where I began. Namely: why can't we get proof for Bigfoot? It's simple: these things are something much more than we assume them to be. History shows that just collecting more and more reports of Bigfoot is getting us nowhere—aside from having to buy more and more filing cabinets. I suggest we'll only get the proof of Bigfoot when we start looking at the mystery from an alternative perspective.

Most investigators of the Bigfoot and cryptid ape phenomena take the view that the beasts they seek are flesh and blood animals that have been incredibly lucky

TIME TRAVEL: THE SCIENCE AND SCIENCE FICTION

in terms of skillfully avoiding us or getting captured and killed. There is, however, another theory that may explain how and why Bigfoot always eludes us, at least when it comes to securing hard evidence of its existence. It's a theory that posits the creatures have the ability to become invisible—that's to say, they can "cloak" themselves so that we do not see them. It's a theory that the bulk of Bigfoot enthusiasts have absolutely no time for. It is a fact, however, that regardless of what people think of the theory, there is certainly no shortage of reports of such a phenomenon. Or perhaps they have the ability to appear to become invisible *but may actually leap into time portals when danger surfaces*. This may sound totally bizarre, but read on.

> **Or perhaps they have the ability to appear to become invisible but may actually leap into time portals when danger surfaces.**

The website *Native Languages* notes: "The Bigfoot figure is common to the folklore of most Northwest Native American tribes. Native American Bigfoot legends usually describe the creatures as around 6–9 feet tall, very strong, hairy, uncivilized, and often foul-smelling, usually living in the woods and often foraging at night.… In some Native stories, Bigfoot may have minor supernatural powers—the ability to turn invisible, for example—but they are always considered physical creatures of the forest, not spirits or ghosts."

Native Americans aren't the only people who hold such beliefs. The website *Bhutan Canada* says: "In 2001, the Bhutanese Government created the Sakteng Wild-

"The Migoi is known for its phenomenal strength and magical powers, such as the ability to become invisible and to walk backwards to fool any trackers."

life Sanctuary, a 253-square-mile protected habitat for the Migoi. The sanctuary is also home to pandas, snow leopards, and tigers but the Bhutanese maintain that the refuge was created specifically for the Migoi. Migoi is the Tibetan word for 'wild man' or more common to Western culture, the Yeti. The Yeti, often called the Abominable Snowman in the west and referred to as the Migoi by the Bhutanese, is a bipedal ape-like creature that is said to inhabit the Himalayan region of Nepal, Tibet, and Bhutan. The Migoi is known for its phenomenal strength and magical powers, such as the ability to become invisible and to walk backwards to fool any trackers."

Davy Russell, who, in 2000, penned an article titled "Invisible Bigfoot," refers to an incident that occurred in 1977 and may be relevant to this particularly charged area of research. The location was North Dakota: "A Bigfoot-type creature was spotted throughout the afternoon and into the evening. Locals, along with the police, staked out the area to search for the mysterious creature. A rancher named Lyle Maxon reported a strange encounter, claiming he was walking in the dark when he plainly heard something nearby breathing heavily, as if from running."

Russell continued that Maxon shone his flashlight in the direction of where the sounds were coming from, but nothing could be seen. Puzzled and disturbed by the encounter, Maxon gave serious thought to the possibility that the beast had the ability to render itself invisible to the human eye.

In April 2012, researcher Mi-Lin said: "This past week, I had several wonderful conversations with a gentleman named Thomas Hughes. Thomas has been communicating with numerous Sasquatch since his first encounter in April 2008. He has a wealth of knowledge about their existence and whereabouts, some of which he shared with me.

"Sasquatch are gentle and playful giants. They range in height from 6–15 feet and live to an age of approximately 120–140 years. They are natural pranksters and are caretakers of Mother Earth. What I mean by caretakers is that they have adapted themselves to the planet instead of trying to change the environment to suit them."

She added that Hughes told her: "They have the ability to raise their frequency just enough to be able to become invisible to humans. They fear humans—seeing them as their greatest threat. So, most of the time, they go invisible when humans are around to avoid being hunted and killed. Sasquatch are aware they are seen from time to time."

On a similar path, *Soul Guidance* offers that Bigfoot is "able to shift the frequency of their physical body, by which it phases out of this physical dimension, and thus enters another dimensional world that lies behind this one...."

"Bigfoot can also shift partially, so they become invisible but are still partially in this physically world. In this partial state, they can follow someone around, invisibly, and their movements can be heard and seen. In this partial state, they can

TIME TRAVEL: THE SCIENCE AND SCIENCE FICTION

walk through walls; and they can sometimes be seen as being transparent, or just the outline of their body. For example, bushes move aside when they step through. When they appear or disappear, their eyes often turn red, probably a characteristic of the shifting frequencies."

Then there is the matter of Bigfoot invisibility in the world of entertainment. From 2000 to 2002, the SyFy Channel ran a series called *The Invisible Man*. It focused on the exploits and adventures of Darien Fawkes (played by Vincent Ventresca), a thief who is recruited into the top-secret world of espionage and assassination. Fawkes is subjected to a form of advanced, fringe surgery that involves the implantation into his brain of something called the "quicksilver gland." It's a gland that has the ability to secrete a substance—quicksilver—that has light-bending abilities when it secretes through the skin, thus creating a form of invisibility.

In series two of *The Invisible Man*, we learn that although the quicksilver gland was being created and synthesized artificially, its development was prompted by the recovery of a Bigfoot corpse and the deduction that its own quicksilver gland was what allowed the creatures to remain almost 100 percent elusive. Fiction paralleling an astonishing truth? It's a question to ponder. Now we get to the crux of the matter.

Ronan Coghlan is the author of a number of acclaimed books, including *A Dictionary of Cryptozoology* and, with Gary Cunningham, *The Mystery Animals of Ireland*. In a 2012 interview with me, Ronan provides his views on the nature of the British Bigfoot and what he believes may be evidence of a link between the phenomenon and that of time portals, UFOs, and more. Having focused his research on what has become known as the British Bigfoot (yes, there are such things; they are seen often), Coghlan says: "The idea that there is a viable, reproducing population of apes or humanoids in Britain is totally risible; it just couldn't be. So, alternative explanations for their presence are to be sought."

He continues: "A lot of the British reports seem to be quite authentic.

"A lot of the British reports seem to be quite authentic," says Ronan Coghlan. "So, there probably are actual beasts or humanoids out there. And the question is: How did they arrive there in the first place?"

So, there probably are actual beasts or humanoids out there. And the question is: How did they arrive there in the first place?" Ronan answers his own question: "It's now becoming acceptable in physics to say there are alternative universes. The main pioneer of this is Professor Michio Kaku, of the City College of New York. He has suggested that not only are there alternate universes, but when ours is about to go out in a couple of billion years, we might have the science to migrate to a more congenial one that isn't going to go out. I think he expects science to keep improving for countless millennia, which is very optimistic of him, but whatever one thinks about that, the idea of alternative universes is now gaining an acceptance among physicists, and he's the name to cite in this area."

The subject is far from one lacking in mysteries and questions, however, as Coghlan acknowledges: "Now, how do you get into, or out of, alternative universes? Well, the answer is quite simple: You have heard of wormholes, I'm sure? No one has ever seen a wormhole, I hasten to add. They are hypothetical, but mainstream physicists say they could be there, and there's one particular type called the Lorentzian Traversable Wormhole. Physicists admit there is a possibility that this exists, and it would be like a short-cut from one universe to another. Thus, for example, it's rather like a portal: Something from the other universe would come through it. Or, something from another planet could come through it."

Turning his attention toward the links between wormholes and bizarre beasts, Ronan comments: "If there are any of these wormholes on Earth, it would be quite easy for anything to come through, and it's quite possible any number of anomalous creatures could find their way through from time to time. You remember John Keel and his window areas? That would tend to indicate there's a wormhole in the vicinity, such as Point Pleasant, West Virginia, where the Mothman was seen.

> *"Now, if UFOs travel by wormholes, and if Bigfoot does the same, that might allow for a connection between the two."*

"I have the distinct suspicion we are dealing with window areas that either contact some other planet, time, or ... universe. My money is on the other universe, rather than the other planet, to be honest with you. Either a short-cut through time, or a short-cut through space, is recognized as possible these days. This is kind of cutting-edge physics, as it were.

"Now, the other one isn't cutting-edge physics at all. It's my own little theory. I think, looking at a great many legends, folk-tales, and things of that nature, it is possible to vibrate at different rates. And if you vibrate at a different rate, you are not seen. You are not tangible. And, then, when your vibration changes, you *are* seen, and you *are* tangible; maybe that this has something to do with Bigfoot appearing and disappearing in a strange fashion.

TIME TRAVEL: THE SCIENCE AND SCIENCE FICTION

"And, finally, on the question of UFOs: Quite a large number of Bigfoot-type creatures have been seen in the vicinity of UFOs. I'm not saying there's necessarily a connection between the two, but they do—quite often—turn up in the same areas. Now, if UFOs travel by wormholes, and if Bigfoot does the same, that might allow for a connection between the two. They might not be mutually exclusive."

CHAPTER 30

THE MATTER OF TELEPORTATION

One of the more intriguing—and highly controversial—claims concerning Area 51 is that top-secret research is undertaken at the base related to none other than teleportation. This issue of teleportation also ties in with the time-traveling saga of the Philadelphia Experiment, described fully in the chapter "From Invisibility to Time Leaps," in which the USS *Eldridge* was reportedly teleported through space and time from Philadelphia, Pennsylvania, to Norfolk, Virginia.

To give some context to that and similar events, let's look a bit deeper at the matter of teleportation. Yes, you did read that right: the very same technology that has become famous in the likes of *Star Trek* and the 1958 movie (and its 1986 remake) *The Fly*. Before we get to the matter of the Area 51 connection to such incredible technology, let's see what teleportation actually is.

The technology company IBM has stated the following concerning this decidedly fringe part of science: "Teleportation is the name given by science fiction writers to the feat of making an object or person disintegrate in one place while a perfect replica appears somewhere else. How this is accomplished is usually not explained in detail, but the general idea seems to be that the original object is scanned in such a way as to extract all the information from it, then this information is transmitted to the receiving location and used to construct the replica, not necessarily from the actual material of the original, but perhaps from atoms of the same kinds, arranged in exactly the same pattern as the original.

"A teleportation machine would be like a fax machine, except that it would work on 3-dimensional objects as well as documents, it would produce an exact copy rather

Teleporation—the feat of disintegrating in one place to reappear in perfect replica in another—may be integral to time travel.

than an approximate facsimile, and it would destroy the original in the process of scanning it. A few science fiction writers consider teleporters that preserve the original, and the plot gets complicated when the original and teleported versions of the same person meet; but the more common kind of teleporter destroys the original, functioning as a super transportation device, not as a perfect replicator of souls and bodies."

With that background from IBM now digested, let's take a look at the rumors concerning Area 51 and teleportation. In 2017 the *Guardian* reported: "Chinese scientists have teleported an object from Earth to a satellite orbiting 300 miles away in space, in a demonstration that has echoes of science fiction. The feat sets a new record for quantum teleportation, an eerie phenomenon in which the complete properties of one particle are instantaneously transferred to another—in effect teleporting it to a distant location."

The story that the *Guardian* referred to concerned a Chinese team that revealed its successes in the field of teleportation in 2017. The BBC ran an article on the astounding story titled "Teleportation: Photon Particles Today, Humans Tomorrow?" It included the following, under the subheading of "What has the Chinese team achieved?":

They created 4,000 pairs of quantum-entangled photons per second at their laboratory in Tibet and fired one of the photons from each pair in a beam

A research team in China achieved "faithful and ultra-long-distance quantum teleportation"—that is, they successfully teleported photons.

of light towards a satellite called Micius, named after an ancient Chinese philosopher. Micius has a sensitive photon receiver that can detect the quantum states of single photons fired from the ground. Their report—published online—says it is the first such link for "faithful and ultra-long-distance quantum teleportation."

"It is a very nice experiment—I would not have expected everything to have worked so fast and so smoothly," says Professor Anton Zeilinger from the University of Vienna, who taught Chinese lead scientist Pan Jianwei.

As for the matter of teleportation in the real world—and possibly at Area 51—we have to turn our attention to a man named Eric W. Davis. In 2004 the U.S. Air Force quietly (as in extremely quietly) contracted Davis's Las Vegas, Nevada–based Warp Drive Metrics company to prepare a report on the feasibility of teleportation. It became known as the Teleportation Physics Study. We know that because the Air Force has now placed the report in the public domain. The specific arm of the Air Force that had a particular interest in teleportation was the Air Force Research Laboratory (AFRL), Air Force Materiel Command, which is based at Wright-Patterson Air Force Base in Dayton, Ohio.

The Air Force states of the AFRL: "Air Force Research Laboratory, with headquarters at Wright-Patterson Air Force Base, Ohio, was created in October 1997.

The laboratory was formed through the consolidation of four former Air Force laboratories and the Air Force Office of Scientific Research. The laboratory employs approximately 10,000 military and civilian personnel. It is responsible for managing and annual $4.4 billion (Fiscal Year 2014) science and technology program that includes both Air Force and customer funded research and development. AFRL investment includes basic research, applied research and advanced technology development in air, space and cyber mission areas.

"With headquarters at Wright-Patterson AFB, Ohio, and an additional research facility at Edwards AFB, Calif., the Aerospace Systems Directorate leads the effort to develop and transition superior technology solutions that enable dominant military aerospace vehicles. Areas of focus include vehicle aerodynamics, flight controls, aerospace propulsion, power, rocket propulsion, aerospace structures, and turbine engines. Programs advance a wide variety of aerospace technologies including unmanned vehicles, space access, advanced fuels, hypersonic vehicles, future strike, and energy management."

It's a little-known fact that the AFRL has an office at Area 51, chiefly because certain sensitive and secret aircraft developed by the brains at the AFRL are tested at Groom Lake—hence the connection. In that sense, a good case can be made that when Davis prepared his report for the AFRL, it almost certainly would have been shared with staff at Area 51. With that said, read on.

The Teleportation Physics Study states: "This study was tasked with the purpose of collecting information describing the teleportation of material objects, providing a description of teleportation as it occurs in physics, its theoretical and experimental status, and a projection of potential applications. The study also consisted of a search for teleportation phenomena occurring naturally or under laboratory conditions that can be assembled into a model describing the conditions required to accomplish the transfer of objects."

> *"… anomalous teleportation has been scientifically investigated and separately documented by the Department of Defense."*

Notably, the document reveals that there was official, secret interest in the field of teleportation that predated the Davis report. On this matter, we have the following from Davis's paper:

"The late Dr. Robert L. Forward stated that modern hard-core SciFi literature, with the exception of the ongoing *Star Trek* franchise, has abandoned using the teleportation concept because writers believe that it has more to do with the realms of parapsychology/paranormal (a.k.a. psychic) and imaginative fantasy than with any realm of science. Beginning in the 1980s developments in quantum theory and general relativity physics have succeeded in pushing the envelope in exploring the reality of teleportation. As for the psychic aspect of teleportation, it became known to Dr. Forward and myself, along with several colleagues both inside and outside of government, *that anomalous teleportation has been scientifically investigated and separately documented by the Department of Defense*" (italics mine).

Davis then provided his own determinations on what specifically constituted teleportation. In his own words:

⊃ *Teleportation—SciFi:* the disembodied transport of persons or inanimate objects across space by advanced (futuristic) technological means. We will call this sf-Teleportation, which will not be considered further in this study.

⊃ *Teleportation—psychic:* the conveyance of persons or inanimate objects by psychic means. We will call this p-Teleportation.

⊃ *Teleportation—engineering the vacuum or spacetime metric:* the conveyance of persons or inanimate objects across space by altering the properties of the spacetime vacuum, or by altering the spacetime metric (geometry). We will call this vm-Teleportation.

⊃ *Teleportation—quantum entanglement:* the disembodied transport of the quantum state of a system and its correlations across space to another system, where system refers to any single or collective particles of matter or energy such as baryons (protons, neutrons, etc.), leptons (electrons, etc.), photons, atoms, ions, etc. We will call this q-Teleportation.

⊃ *Teleportation—exotic:* the conveyance of persons or inanimate objects by transport through extra space dimensions or parallel universes. We will call this e-Teleportation.

Davis suggested that the "p-Teleportation" would be the most profitable phenomenon of all: "P-Teleportation, if verified, would represent a phenomenon that could offer potential high-payoff military, intelligence and commercial applications. This phenomenon could generate a dramatic revolution in technology, which would result from a dramatic paradigm shift in science. Anomalies are the key to all paradigm shifts!"

In a portion of the report titled "Recommendations," Davis noted: "A research program ... should be conducted in order to generate p-Teleportation phenomenon in the lab. An experimental program ... should be funded at \$900,000–1,000,000 per year in parallel with a theoretical program funded at \$500,000 per year for an initial five-year duration."

The official line, when the media got wind of the story, is that the Air Force hastily discontinued the program.

CHAPTER 31

FROM DEJA VU TO THE MATRIX

As we have seen, there's very little that we know for sure about time travel and those who travel through time. To come at least somewhat closer to understanding the phenomenon, there are two issues that need to be addressed related to the topic. One is the experience of déjà vu, and the other is the possibility that we are all living in a real-life world hidden by an illusory one, akin to the one in the movie *The Matrix*. Both of them, in strange ways, impact on the matter of what can accurately be titled "time anomalies."

There can be very few people, if any at all, who have never experienced the phenomenon of déjà vu, a term that means "already seen." There are two overriding theories for what causes déjà vu. One is that now and again, time itself has a glitch, albeit usually a very brief one that leaves us feeling slightly weird and a bit uneasy. The other theory is far more down to earth: that déjà vu is caused by the occasional misfiring of the brain.

Crystal Raypole at *Healthline* has addressed the matter and provides us some fascinating material: "'Déjà vu' describes the uncanny sensation that you've already experienced something, even when you know you never have. Say you go paddleboarding for the first time. You've never done anything like it, but you suddenly have a distinct memory of making the same arm motions, under the same blue sky, with the same waves lapping at your feet. Or perhaps you're exploring a new city for the first time and all at once feel as if you've walked down that exact tree-lined footpath before. You might feel a little disoriented and wonder what's going on, especially if you're experiencing déjà vu for the first time."

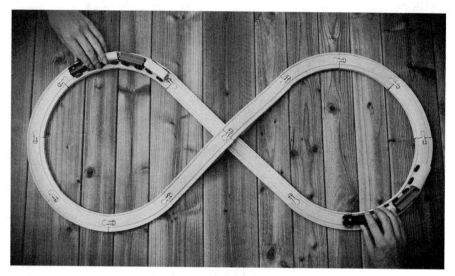

Could the phenomenon of déjà vu be evidence of a glitch in time?

The popular magazine *Scientific American* has tackled this slightly jarring phenomenon, too: "Carrie-Anne Moss, as Trinity in *The Matrix* trilogy, tells us (and Keanu Reeves as Neo) that déjà vu is a 'glitch in the Matrix'—the simulated reality that keeps humanity unaware that intelligent machines have actually taken over the world. That explanation is perfect for cyberpunk science fiction, but it doesn't give us any scientific understanding of the phenomenon. We associate the feeling of déjà vu with mystery and even the paranormal because it is fleeting and usually unexpected. The very things that intrigue us about déjà vu are the same things that make it hard to study. But scientists have tried using tricks like hypnosis and virtual reality."

Writers Julia C. Teale and Akira R. O'Connor at *Frontiers for Young Minds* offer these words: "The percentage of people who experience déjà vu is probably somewhere between 30% (about 8 in a class of 30) and 100% (everyone in a class of 30). We are not sure about the exact percentage for two important reasons. First, we cannot ask everyone in the world so we have to use the results of surveys of small groups of people. This is a problem because surveys can give us quite different results depending on who we ask. Second, people can give very different answers depending on the definition we give of déjà vu. Asking the question in different ways can get very different results.

"We can also get an idea of how often déjà vu happens by asking people. Again, the answers they give depend on who they are and how we ask them the question, but most people report déjà vu somewhere between every few weeks and every few months. Typically, this means that déjà vu is not very common so if you have experienced it recently you are very lucky!"

Going back to the matter of the mind itself causing déjà vu, the *Penn Medicine* blog of the University of Pennsylvania offers some interesting data, citing Dr. Roderick C. Spears, a physician with Penn Neurology Valley Forge: "Though much rarer, déjà vu is sometimes a sign of a seizure, specifically an epileptic seizure. 'About 60 percent of people with epilepsy have something called a focal seizure, which is in just one part of the brain. This can be in the same part of the brain where memory is stored: the temporal lobe,' says Dr. Spears. Focal seizures can be hard to recognize as seizures because they are short and you remain conscious throughout. A person having one may look like they are having a staring spell or daydreaming.

"Some people who have focal seizures may experience intense feelings of déjà vu. This is often accompanied by:

⊃ Motor feelings, which impact your ability to control your muscles, like twitching or the inability to move certain muscles;

⊃ Sensory feelings, involving taste, touch, smell, vision, and hearing, like tasting or hearing something that's not real;

⊃ Sudden and unexplainable feelings of joy, anger, sadness, or nausea;

⊃ Strange, repetitive behaviors, such as blinking, twitching, or moving your mouth involuntarily;

⊃ An unusual sensation that a seizure is about to occur, called an aura."

Aside from the medical explanations for déjà vu, what about the movie-themed explanation for the mystery? Did the *Matrix* movies inadvertently play into a real situation—that reality and time are not what they appear to be? Let's look a bit more at these huge cinematic successes.

In the 1999 motion picture *The Matrix*, starring Keanu Reeves, Carrie-Anne Moss, and Laurence Fishburne, we are introduced to a world that appears identical to the one in which we all live but that, in "reality," is nothing of the sort. In the film, the rise of artificial intelligence in the twenty-first century led the human race to go to war against increasingly powerful machines and computers that had no wish to play second to humans. Unfortunately, humans lost the war, and the machines won. As a result, the human race is kept in check in a very strange and disturbing way. We are no longer born; instead, we are *grown* in vast factories. The machines "feed" on our bioelectricity, using it as fuel. But how do they prevent us from rising up against them? Here is where we get to the crux of the story—and to the title of the movie franchise.

TIME TRAVEL: THE SCIENCE AND SCIENCE FICTION

*Did the Matrix movies inadvertently play into a real
situation—that reality and time are not what they
appear to be?*

The world we assume we live in is nothing but an infinitely advanced computer-based simulation. It is something akin to a sophisticated dream—a vast online dream that we are all unknowingly hooked into. This simulation is called the Matrix. Our real, physical lives, from birth to death, are spent endlessly sleeping in large pods. They are pods in which we are kept alive and fed, ensuring nourishment for the machines. Occasionally, a glitch occurs in the Matrix, which can provoke the likes of time-muddling déjà vu in our unreal environment. Even more infrequently, someone will break free of the chains of the Matrix, fight against their deadly controllers, and learn the shocking truth, which is exactly what happens across the course of the *Matrix* trilogy. In the movies, Keanu Reeves's character, Thomas Anderson (aka Neo), believes that he lives in the latter part of the twentieth century. He doesn't. As we learn, it's actually 2199, or thereabouts. The real date has been lost. We therefore have the phenomenon of déjà vu and a world in which time is, in essence, a dream. No wonder I decided to bring this matter to your attention, right? Right! And matters don't end there. How about a real-life equivalent of one of the most memorable moments in *The Matrix*?

The so-called contactees were people who, chiefly in the 1950s, claimed encounters with very human-looking aliens who demanded that we humans lay down our atomic weapons. Among the contactees were those controversial characters who claimed close encounters with the aliens that became known as the "Space Brothers." One of the contactees, whose story fits right into this subject of mind alteration, was Orfeo Angelucci. In correspondence with Jim Moseley of *Saucer Smear*, Angelucci said he had been visited in 1954 by people from both the FBI and U.S. Army Intelligence. That was not at all surprising, as most of the contactees had files opened on them—largely because of their politics rather than their alien claims. One of the stories that Angelucci shared with Moseley was particularly strange. On one particular night in December 1954, after finishing working out in Twentynine Palms, California, Angelucci headed out to a local diner. That's where things got strange. Angelucci recorded: "I felt a strangeness in the air. There is a cosmic spell over the desert most of the time, but tonight the mystery was less distant and intangible; it was close and pulsating."

Angelucci was soon deep in conversation in that diner with a man who identified himself only as Adam, a customer who claimed to be 30-something and suffering from a terminal illness. Death was said to be just around the corner for the man. In an odd and synchronistic fashion, Adam claimed that he had read Angelucci's book *The Secret of the Saucers*, that he considered their meeting to be beyond just an amazing coincidence, and that he wished to share his thoughts with Angelucci before time ran out. That is, he wanted to share them quite literally. But, said Adam, before their conversation could begin, Angelucci had to swallow a pill. Angelucci didn't know what kind of a pill this was, but that didn't stop him from doing exactly what Adam demanded from him. Angelucci swallowed the "oyster-white pellet." For Angelucci, there was now no turning back. It didn't take long before he felt odd and out of this world. Spaced out. Fucked up. In short, Angelucci had been drugged. It was almost like one of the most famous scenes of *The Matrix*. You know the scene: the red pill versus the blue pill. But this was the world of the real, not of Hollywood.

Angelucci said: "I took the pellet and dropped it into my glass. Immediately the water bubbled, turning slowly into the clear, pale amber contained in [Adam's] own glass. I lifted the glass a few inches from the table, looking into it with a feeling that this might be the drink I dared not hope for. The exhilarating aroma rising from it could not be mistaken.... I thrilled from head to foot as I took the glass, lifted it to my lips, and swallowed twice from it. At that instant, I entered, with Adam, into a more exalted state and everything around me took on a different semblance. No longer was I in Tiny's café in Twentynine Palms. It had been transformed into a cozy retreat on some radiant star system. Though everything remained in its same position, added beauty and meaning were given to the things and people present there."

Almost as an aside, Angelucci said: "Among the patrons dining that evening were two marines from the nearby base. Sometimes they glanced our way as they talked and drank beer following their meal." Angelucci said that Adam seemed oddly obsessed with the glass and was "fraught with expectancy." Suddenly, the sounds of music filled Angelucci's ears. Incredibly, the music seemed to be coming from the glass itself, or so it seemed to Angelucci. The reality was that he was now completely and utterly stoned.

Angelucci stared at the glass and saw the figure of "a miniature young woman" dancing inside the glass! That's right: the drugs were now kicking in to a high degree. Of the small woman, Angelucci said: "Her golden-blond beauty was as arresting as the miracle of her projection in the glass. Her arms moved in rhythmic motion with the graceful thrusts of her dancing body." What began as a pleasant meeting between like-minded souls soon became a drug-driven interrogation. By Angelucci's own admission, he spilled the entire beans to Adam: the nature of his encounters and the words of his alien friends. There was even a debate on politics, which is rather telling. Angelucci staggered home, his mind hardly his own for the next few hours. It's important to note that there is much more to all of this. Read on.

Why were the Space Brothers so concerned that we would destroy ourselves in the 1950s? At first glance, the answer is very simple: they liked us and wanted us to stay alive! Maybe, however, there was more to it than that. Perhaps there were even disturbing reasons for that apparent concern for our welfare—and even our existence.

Before we get to the heart of it all, however, let us first take a look at the Space Brothers, for those who may not be acquainted with the strange subject. Although it was the summer of 1947 when the term "flying saucer" was coined, sightings of—or encounters with—alleged aliens didn't really begin on a large scale until the early 1950s. That's when the aforementioned Space Brothers surfaced from wherever they came. It also happens to be the period in which the matter of nukes began to surface.

If the Space Brothers were actually time travelers trying to manipulate the present to alter the future—and to ensure that a third world war didn't erupt and destroy civilization—then matters become much more understandable.

The Space Brothers were described as looking eerily human-like. The major difference was that the males had very long hair, which, of course, was a rarity during the dawning of the 1950s. The women looked like women on Earth. Both the males and the females occasionally had some very slight differences in their facial appearance, but nothing that really stood out as odd or unusual. The aliens chose certain figures to spread the word that the human race should get rid of its nuclear weapons—and that if we didn't follow the path of the creatures from other worlds, then we would surely all be fried in a radioactive holocaust of our very own making. Those with whom the Space Brothers and Space Sisters chose to work became known as the contactees. The very long list included George Van Tassel, Dana How-

The long-haired Space Brothers of the 1950s may have been time travelers from the future trying to prevent a third world war.

TIME TRAVEL: THE SCIENCE AND SCIENCE FICTION

ard, Truman Bethurum, Mollie Thompson, Orfeo Angelucci, George King, and Margit Mustapa. And they were just the tip of the iceberg.

The beings from faraway planets would often meet the contactees late at night or in the early hours of the morning, and very often in out-of-the-way places, such as deserts, woods, mountains, or below old bridges. As the 1950s progressed and our nuclear arsenals grew, so did the aliens' concerns for the human race. But was that apparent concern really due to the benevolent, friendly nature of the Space Brothers? Maybe not. What if the Space Brothers were not from another solar system after all? What if they were from right here on planet Earth, but from our future? After all, why would a space-faring race from a faraway solar system care about us, a civilization countless light-years away? The answer is quite simple: they wouldn't care.

If, though, the Space Brothers were actually time travelers trying to manipulate the present to alter the future—and to ensure that a third world war didn't erupt and destroy civilization—then matters become not just clearer but much more understandable.

Déjà vu–driven time glitches, a blockbuster movie that just might reflect the true reality of our world, a story that takes place around 2199 but that is believed to be 1999, and a real example of the famous "blue pill or red pill" scene in *The Matrix*—all of it suggests that science fiction may be closer to science fact.

TIME TRAVEL: THE SCIENCE AND SCIENCE FICTION

all: Stonehenge. It was in the fall of 2002 when a man named George Price, at the time serving with the British Army, had a bizarre encounter on a huge expanse of English wilderness called Salisbury Plain, which is home to Stonehenge.

It was on the plain that Price, taking part in army training maneuvers at the time, caught sight of an immense creature of apelike proportions and appearance and sporting a coat similar to that of an orangutan. As the military closed in, the beast wasted no time in exiting the area at high speed and vanishing into the undergrowth. That the monster should have been encountered near one of the world's most famous sacred sites—one that dates back to approximately 3100 B.C.E.—is surely no coincidence.

Late one night, on the remote Scottish island of Hoy in the early 1940s, Thorner was stunned by a bizarre sight: a band of hairy people engaged in a wild dance near the edge of a large cliff.

Then there is the curious tale of a man named W. E. Thorner. Late one night, on the remote Scottish island of Hoy in the early 1940s, Thorner was stunned by a bizarre sight: a band of hairy people engaged in a wild dance near the edge of a large cliff. Thorner said: "These creatures were small in stature, but they did not have long noses nor did they appear kindly in demeanour. They possessed round faces, sallow in complexion, with long, dark, bedraggled hair. As they danced about, seeming to throw themselves over the cliff edge, I felt that I was a witness to some ritual dance of a tribe of primitive men. It is difficult to describe in a few words my

At Kit's Coty House—a megalith in Kent, England, built between 4500 and 3800 B.C.E.—a large figure wearing a loincloth and furred boots, perhaps a Neolithic hunter, suddenly appeared before a modern-day visitor.

TIME TRAVEL: THE SCIENCE AND SCIENCE FICTION

feelings at this juncture or my bewilderment. The whole sequence could have lasted about three minutes until I was able to leave the cliff edge."

Still on the topic of unknown, hairy, humanoid animals seen in the United Kingdom—and ones seen in the direct vicinity of ancient, historical sites—crypto-zoologist Neil Arnold told me the following: "I've always wondered what type of manifestation these U.K. 'wild men' could be. In the autumn of 2011 a psychic lady who I know as a friend and who I trust—I don't often have any interests in psychics—accompanied me to Blue Bell Hill, which is a very haunted village in Kent, a few miles short of the town of Maidstone. I knew of several obscure 'man-beast' reports in the area which she knew nothing about. I took her to one particular spot, near some ancient stones, hoping she'd pick up a ghostly presence and she said she felt nothing whatsoever, but she did state quite categorically that a few years previous, around 2003, she'd had a bizarre encounter in the area one night."

Arnold continued: "She had visited Kit's Coty House—a set of stones—with a group of fellow psychics. Her friends were over on one side of the field which harbors the stones and she was in another area when she noticed someone walking towards her a few hundred yards away. The figure seemed to be striding rather aggressively and was coming from the direction of a thicket which runs alongside the field."

The woman, whose name is Corriene, said Arnold, "stated that from a distance the figure appeared huge in build and covered in hair and she sensed it was not 'real' but gave off an air of malevolence. The figure marched towards her and she could see it had long hair and a beard, covering most of its face. The hulking figure was taller than six feet and appeared to have a loin cloth around its waist and furred boots. No one else saw this figure, but I was intrigued as I knew that in the past several witnesses had come forward to say they'd seen similar figures in woods within miles of Blue Bell Hill."

"Blue Bell Hill and much of Kent is steeped in history—so maybe people were seeing some type of Neolithic hunter."

Arnold told me this set him on an intriguing and alternative pathway: "I began to wonder if people had seen, from a distance, some type of ghostly primitive man—long hair, bearded, muscular, animal fur around the waist—who, from several hundred yards away, or in ill light, may have looked as if he was covered in hair. Blue Bell Hill and much of Kent is steeped in history—so maybe people were seeing some type of Neolithic hunter. Corriene was intrigued by what I said and then, rather startled, mentioned that on another occasion whilst in the area of the stones she'd seen several of these people who she felt were not aggressive, and although armed with spears were simply guarding the area and stooping low in the bushes, curious as to what they were seeing."

All of these creatures are out of place. More importantly, they are clearly all out of time, too.

TIME TRAVEL: THE SCIENCE AND SCIENCE FICTION

CHAPTER 33

EXAMINING THE II:II ENIGMA

Perhaps related to the issue of time travel is the very strange subject of what has become known as the "11:11 phenomenon." In essence, people all over the world have experienced strange and near-supernatural experiences at 11:11, both in the morning and in the evening. And the more people become aware of this puzzle, the more likely they are to get further encounters with these by now legendary numbers.

What exactly is the 11:11 phenomenon? It relates to how times can affect and manipulate our actions. Why does this happen? Let's see what a number of experts in this field have to say about all of this.

Marie Jones and Larry Flaxman are the authors of *11:11 The Time Prompt Phenomenon: Mysterious Signs, Sequences, and Synchronicities*. It's a book that gets right to the heart of the mystery. They write: "Do you wake up every night and see 11:11 on the clock? Or 3:33? 4:44? Does the same number sequence seem to appear throughout your life over and over? Did you know that millions of people all over the world experience the same phenomenon? These mysterious number sequences are known as 'time prompts,' and show up on digital clocks, cell phones, receipts, billboards, advertisements, and other places. They seem like pure coincidence, but what if they are actually messages from a higher source, like angels, guides, or even the Universe itself, urging you to pay attention to something important?"

The people at the website *Power of Positivity* say: "This phenomenon basically occurs to remind one of the profound synchronicities and cosmic shifts occurring

The 11:11 phenomenon relates to how times can affect and manipulate our actions.

during this beautiful time on the planet. Also, it brings your attention to your present thoughts and feelings. Your angels' underlying intention bringing our awareness to 11:11 is to make us more conscious of ourselves. Plus, it serves to remind each of us that we always have guidance and greater wisdom when we feel stuck, scared, or frustrated…. According to Doreen Virtue, a well-known angel therapist and psychic medium between the spirit world and the human world, seeing repeating numbers, especially these angel numbers, 11:11, means you should focus on keeping your thoughts positive. That's because your desires will manifest instantly into form. Put all your attention on what you desire instead of what you fear. And soon, your angels will continue to reward you."

Kate Rose, writing for *Elephant Journal,* has some ideas as to what's going on. She writes: "11:11 is the universe's way of urging us to pay attention to our heart, our soul and our inner intuition. It's serving as a wake-up call to us so that opportunities are not missed in this lifetime. Seeing this particular sequence means that the universe is trying to have us open our eyes and begin paying more attention to the synchronicities around us.

"Ever think about someone and suddenly they call or text? Or you see their name places, or meet people by the same name? Do you physically run into someone that has been running through your conscious or unconscious mind? These are not coincidences, but synchronicities. *The Universe tries to show us the way through small signs, but the trick is we have to be ready to listen to them.*

"Seeing 11:11 also is a sign of experiencing accelerated soul growth, which means that we may soon be finding ourselves living the life we had previously only thought about. Our inner world is changing and we may find people and events coming unexpectedly into our lives—but at just the right time."

At a UFO conference in 2016, I spoke with a young man who had developed an obsession with the 11:11 phenomenon and its potential ties to the UFO phenomenon. Everywhere he went, he told me, he saw the number 11:11, whether on his watch, on his laptop, or even on his microwave. I pointed out to him that, rather ironically, I, too, had had a few such experiences of the 11:11 kind. But I also told him that if you are *aware* of the 11:11 phenomenon, then it's nearly inevitable you're going to take extra notice of those specific figures—and *remember* when you see them. With that in mind, I asked him something broadly along these lines: "How often do you take notice, or remember, when you see the numbers 10:45 or 08:36?" He had no answer. Of course he didn't, because those numbers had no meaning to him. Why would they?

My suggestion that he get out a bit more and make some friends—which might have taken his mind off sitting alone and constantly waiting for 11:11 to come around every 12 hours—was not what he wanted to hear.

The story is not quite over. He had brought with him to the gig a thick, leather-bound journal packed with endless scribbling concerning the 11:11 issue and that spanned, if I recall correctly, a period of close to eight or nine months. Over the 20 or so minutes that we chatted, it became disturbingly clear to me that this man's life was being directed by (as he saw it) invisible entities that lurked in his home and endlessly tormented and manipulated him with that now-ubiquitous number. My suggestion that he get out a bit more and make some friends—which might have taken his mind off sitting alone and constantly waiting for 11:11 to come around every 12 hours—was not what he wanted to hear. It was, however, absolutely all I had to offer.

One of the weirdest things about all of this: literally at the time of writing this portion of the book, I experienced several 11:11 events.

I could go on with many more such examples, but I won't, because you get the picture. In conclusion, I should stress (*and stress again*, for those who may misinterpret my words) that I'm not criticizing people who follow the 11:11 issue. Nor am I criticizing those who feel that certain UFO, time travel, or other significant events occur on specific dates because of a specific reason. And I'm not even criticizing ufology as a body, unified or semifractured. All I'm doing is pointing out that when UFOs, time anomalies, or related phenomena become the *dominating* factor in a person's life, it's not a good thing. By all means, investigate the strange phenomena and chase down significant cases. Do some fieldwork, if that's your thing, and maybe some blogging and article writing, too. But don't do it at the expense of *living*.

One of the weirdest things about all of this: literally at the time of writing this portion of the book, I experienced several 11:11 events. Was I being pushed on to write the book? Perhaps, given that the theme of the book you are now reading is time itself.

TIME TRAVEL: THE SCIENCE AND SCIENCE FICTION

Now we come to the matter of seeing into the future, but in just about the worst way possible.

One of those who lost their lives on that catastrophic morning was Eryl Mai Jones. At the time, she was just 10 years of age and lived in the village. On October 19—two days before so much death engulfed Aberfan and its people—Eryl told her mother she had had a terrifying dream. Or, rather, an absolute nightmare. Patrick Lynch of *History Collection* says of this unsettling aspect of the story: "The 10-year-old told her mother that in her dream, she had gone to school only to discover that it was gone because something black was covering it. It was the latest in a week's worth of unusual behavior from Eryl. In the days leading up to the disaster, she told her mother that she wasn't afraid to die because she would be 'with Peter and June.' Those were the names of two former schoolmates who had died young."

Cases like this one are particularly confounding because, like so many of the examples presented in the pages of this book, they demonstrate that time is not what it appears to be.

Lynch continues: "Tragically, Eryl was proven correct but her life, and the lives of 143 others could have been saved if the NCB [National Coal Board] had paid attention to the complaints about the spoil tip that caused the disaster. In 1963, Eryl's school, Pantglas, sent a petition to the NCB which complained about the danger of the tip. Although every mining community had tips, this particular one was a problem because it lay on porous sandstone with streams and underwater springs. It had slipped in 1965, but no one was hurt. The NCB didn't want to investigate the problem and basically suggested that if the town made a fuss, the mine would be closed and that would be an economic catastrophe."

How did poor little Eryl come to have that brief glimpse of the disaster that would take her life, two days later, and in such an eerie fashion? Indeed, it's very difficult, if not impossible, to dismiss what Eryl said: that the school was missing "because something black was covering it."

Cases like this one are particularly confounding because, like so many of the examples presented in the pages of this book, they demonstrate that time is not what it appears to be, that we can see into the future and encounter things that have yet to be seen. And, as this case also demonstrates, to see into the future doesn't necessarily require a high-tech time machine. The mind itself is able to see ahead.

CHAPTER 35

A BRIEF WALK AROUND VERSAILLES

N ow we come to one of the most controversial, and most enduring, tales of time travel. It's a story that dates back to 1901. Author and anomalies researcher Micah Hanks says: "In 1901, two women claimed that while visiting the Palace of Versailles together, a number of strange phenomena began to occur which led them to feel that space (and perhaps time) had changed in a very unconventional way. Known today as the Moberly-Jourdain incident, the story remains among the more famous cases of purported time travel that occurs without the aid of any supplemental machinery."

Genevieve Carlton wrote of this incredible story: "A century ago, a Versailles time-traveling incident made headlines around the world. Two academics named Charlotte Anne Moberly and Eleanor Jourdain experienced a strange event during a casual stroll through the gardens of Versailles in 1901. The women spent the next decade researching every detail from their trip, finally concluding that they had experienced a time slip. Since then, many have considered Moberly and Jourdain as real-life time travelers.

"Moberly and Jourdain's book, republished as *An Adventure: A True Story about Time Travel*, recounts every detail of the incident, including exhaustive research into the clothing, layout, and people at Versailles in 1789. Some of the details they recall simply don't appear in most history books. And with all the fantastic stories about Versailles, it's not hard to imagine that mysterious things still might occur at the palace."

TIME TRAVEL: THE SCIENCE AND SCIENCE FICTION

Let's ponder on the following from Kathleen McGowan: "On 10 August 1901, Moberly and Jourdain toured the palace of Versailles, which apparently neither of them were really impressed with (this becomes potentially important later as a theme) but they got lost along the way to the smaller Trianon Palace in the gardens. Now, this is not hard to do, and whereas I have personally been to Versailles many times, I still get lost in the miles—yes, miles—of gardens. The tree lined avenues are long and confusing in the way they intersect and lead to different areas of the main and smaller palaces. And there is a strange, disconcerting feeling to the gardens at certain times of the day which I cannot fully explain. But I will say that I have had emotional—to the point of erratic—overload while walking those avenues myself, so I think that naturally sensitive people are prone to the inherent mysteries of Versailles."

Interestingly, Jourdain experienced something very similar. In her own words: "I began to feel as if I were walking in my sleep; the heavy dreaminess was oppressive. At last we came upon a path crossing ours, and saw in front of us a building consisting of some columns roofed in, and set back in the trees. Seated on the steps was a man with a heavy black cloak round his shoulders, and wearing a slouch hat. At that moment the eerie feeling which had begun in the garden culminated in a definite impression of something uncanny and fear-inspiring. The man slowly turned his face, which was marked by smallpox: his complexion was very dark. The expression was very evil and yet unseeing.... I felt a repugnance to going past him."

Back to the tale of Moberly and Jourdain, Francky Knapp expanded on an admittedly creepy part of the story: "The women reached the *Temple de l'Amour* when

Charlotte Anne Moberly and Eleanor Jourdain were strolling through the gardens of Versailles when they encountered several historical figures, including Marie Antoinette.

Moberly locked eyes with a harrowing figure. That's when she began to (understandably) lose her cool. 'The expression [of the man] was evil yet unseeing,' she said, recalling his complexion as pockmarked, and that 'although I did not feel that he was looking particularly at us, I felt a repugnance....'

"Luckily, another man of a more dapper disposition appeared, and kindly pointed them towards the *Petit Trianon*, where Moberly saw a 'fair-haired woman' in a summer dress sketching on the lawn. The woman, Moberly would later deduce, was none other than Marie-Antoinette; her appearance was identical to that of a 1785 portrait of the queen by Adolf Wermüller, and which Moberly came across haphazardly after the incident."

It's clearly not just a story of two women briefly traveling into the past. Rather, the whole story is one saturated in creepiness—even a sense of paranormal malignance.

Offering a more prosaic explanation, historian Philippe Jullian suggested that the two women merely happened upon a fancy dress party and saw the attendees dressed in period costumes. Leila McNeill, writing in *Mental Floss*, addresses this theory and the latter days of the two women: "Dame Joan Evans, a historian and former student at St. Hugh's, obtained the copyright to *The Adventure* as Jourdain's literary executor and accepted Jullian's explanation of the events, suspending printing after the fifth edition in 1955. Moberly and Jourdain, however, never retracted their stories. In 1924, Jourdain became embroiled in a scandal at St. Hugh's after wrongfully firing a tutor; it was clear she would be asked to resign by the college council, but died of heart failure at the age of 61 before they could do so. Moberly died in 1937 at the age of 90, still telling the story of her adventure at Versailles to those who would listen."

There's no doubt at all that this particular story is both fascinating and weird. I say "weird" because it's clearly not just a story of two women briefly traveling into the past. Rather, the whole story is one saturated in creepiness—even a sense of paranormal malignance. Why that should be the case is something we are unlikely to ever know.

CHAPTER 36

THE WOMEN IN BLACK

Within the world of UFO research, the Men in Black are about as legendary as they are feared. These pale-faced, ghoulish entities have for decades terrorized into silence both witnesses to and researchers of UFO encounters. Theories on who or what the MIB might be are legion. They include extraterrestrials, government agents, demonic creatures, vampires, time travelers from the future, and interdimensional beings from realms that coexist with ours. There may very well be more than one explanation for the unsettling phenomenon. The explanation offered by Joshua P. Warren, among others, is that the MIB are time travelers.

While much has been written on the sinister and occasionally deadly actions of the MIB, very little has been penned on the subject of their equally bone-chilling companions: the *Women* in Black. Make no mistake: the WIB are all too real. And they are as ominous, predatory, and dangerous as their male counterparts—and as likely to traverse time. In the same way that the Men in Black don't always wear black but sometimes wear military uniforms or beige-colored outfits, so do the WIB often don another form of dress, particularly white costumes. In that sense, the term "WIB" is, like "MIB," one that is somewhat flexible in terms of actual nature and description.

The WIB may not have achieved the iconic status of the MIB, at least until recently, but these fearsome females, and their collective role in silencing those who immerse themselves in the UFO puzzle, the domains of the occult, and the world of the paranormal, are terrifyingly real. Not only that; the WIB have a long and twisted history.

Years before they plagued and tormented flying saucer seekers, the Women in Black roamed the landscape by night, stealing babies and young children and plaguing the good folk of the nineteenth-century United States and United Kingdom. They were also up to their infernal tricks in the 1920s.

Years before they plagued and tormented flying saucer seekers, the Women in Black roamed the landscape by night, stealing babies and young children and plaguing the good folk of the nineteenth-century United States and United Kingdom.

A definitive WIB surfaced in nothing less than a piece of publicity-based footage for a Charlie Chaplin movie, *The Circus*, which was made in 1928. The footage, undeniably genuine and shown not to have been tampered with, reveals what appears to be a short, elderly lady wearing a long black coat and a black hat pulled low over her face, walking through Los Angeles in the West Coast heat. If that was not strange enough, she is clearly holding to her ear what appears to be a cell phone and is talking into it as she walks. Weirder still, the Woman in Black sports an enormous pair of black shoes, which look most out of place given her short stature. She also seems to be taking careful steps to avoid allowing her face to be seen clearly. Might she have been some kind of time-traveling Woman in Black, working hard—but spectacularly failing—to blend in with the people of Los Angeles all those years ago?

Fifteen years later, a terrifying WIB haunted the Bender family of Bridgeport, Connecticut. It so happens that a certain Albert Bender of that clan near-single-handedly began the Men in Black mystery. In the early 1950s, Bender, after establishing the International Flying Saucer Bureau, was visited and threatened with nothing less than death by a trio of pale, skinny, fedora-wearing MIB. These visits firmly set the scene for the decades of MIB-themed horror and mayhem that followed. Bender's visitors were not secret agents of the government, however. He said they materialized in his bedroom—a converted attic in a creepy old house of *Psycho* proportions—amid an overpowering stench of sulfur. They were shadowy beings with demon-like, glowing eyes. We surely cannot blame the CIA, the FBI, or even the all-powerful NSA for that!

In 1956 UFO sleuth Gray Barker penned a book on Bender's confrontations with the Men in Black. It was titled *They Knew Too Much about Flying Saucers* and became a classic. Six years later, Bender penned his own book on his encounters with the MIB: *Flying Saucers and the Three Men*. It was these two books that brought the MIB into the minds and homes of flying saucer enthusiasts across the world. After this, Bender dropped each and every one of his ties to ufology. He was careful to avoid speaking about the subject ever again and thereafter focused his time on running the appreciation society of composer Max Steiner.

Before this, however, back in the 1930s, the Bender family had a black-garbed woman in its midst who tormented both young and old in the dead of night. Predating Albert Bender's own experience with the MIB by years, the hideous silencer in black haunted the Benders almost endlessly. For the Bender family, long before the Men in Black, there was a Woman in Black.

For the Bender family, long before the Men in Black,
there was a Woman in Black.

In the 1960s, the emotionless, evil-eyed WIB turned up in the small, doom-filled town of Point Pleasant, West Virginia. It was right around the time that sightings of the legendary flying monster known as Mothman were at their height. Claiming to be "census takers," these pale-faced, staring-eyed WIB practically forced their way into the homes of frightened witnesses to Mothman. What began as seemingly normal questions about the number of people in the house, of the average income of the family, and of the number of rooms in the relevant property soon mutated into something much stranger: persistent and intrusive questions about strange dreams, about unusual telephone interference, and about beliefs regarding the world of all things of a paranormal nature.

One of the WIB that put in an appearance at Point Pleasant claimed to have been the secretary of acclaimed author on all things paranormal John Keel, author of *The Mothman Prophecies*. Just like her male counterparts, she turned up on doorsteps late at night, waiting to be invited in, before grilling mystified and scared souls about

One Woman in Black, claiming be the secretary of the paranormal expert John Keel, turned up on people's doorsteps late at night to grill the mystified souls about their UFO and Mothman encounters before vanishing back into the night.

TIME TRAVEL: THE SCIENCE AND SCIENCE FICTION

their UFO and Mothman encounters. She would then vanish into the night after carefully instilling feelings of distinct fear in the interviewees. Only when dozens of such stories got back to Keel did he realize the sheer, incredible scale of the dark ruse that was afoot. Keel had to break the unsettling news to each and every one of the frightened souls who contacted him: "*I have no secretary.*"

In the 1970s, wig-wearing and anemic-looking WIB made life hell for more than a few people who were unfortunate enough to cross their paths. Something similar occurred in England, Scotland, and Ireland during the 1980s: a weird wave of encounters with "phantom social workers" hit the United Kingdom. They were out-of-the-blue encounters that eerily paralleled the incidents involving WIB-based "census takers" of West Virginia back in the 1960s.

Just as menacing, sinister, and unsettling as their American cousins, these particular WIB began by claiming that reports had reached them of physical abuse to children in the family home that had to be investigated. Worried parents, clearly realizing that these haglike crones were anything but social workers, invariably phoned the police. The WIB, realizing when they had been rumbled, made hasty exits, always before law-enforcement personnel arrived on the scene. Most disturbing of all, there was a near-unanimous belief on the part of the parents that the Women in Black were intent on kidnapping the children for purposes unknown, but surely no good.

Back to the United States, in early 1987, Bruce Lee, a book editor for Morrow, had an experience with a WIB-type character in an uptown New York bookstore. Lee's attention to the curious woman—short, wrapped in a wool hat and a long scarf, and wearing large black sunglasses behind which could be seen huge, "mad dog" eyes—was prompted by something strange and synchronistic. She and her odd partner were speed-reading the pages of the just-published UFO-themed book *Communion* by Whitley Strieber. It was a book published by the very company Lee was working for. Lee quickly exited the store, shaken to the core by the appearance and hostile air that the peculiar pair oozed in his presence.

> *They dwell within darkness, they surface when the landscape is black and shadowy, and they spread terror and negativity wherever they walk.*

In 2001 Colin Perks, a British man obsessed with finding the final resting place of King Arthur, received a nighttime visit from a beautiful but emotionless Woman in Black who had near-milk-white skin. She claimed to represent a secret arm of the British government that was intent on shutting down research into all realms of the paranormal. When Perks defiantly and defensively said he would not be stopped by veiled threats, the Woman in Black responded with a slight, emotion-free smile and advised him he had just made a big mistake and that he should soon expect another visitor. That other visitor soon turned up, late one dark night. It was a hideous, gargoyle-like beast with fiery, blazing-red eyes that loomed large over Perks's bed in the early hours. Perks, forever thereafter blighted by fear and paranoia, came to believe his Woman in Black and the winged beast were one and the same: namely, a monstrous shape-shifter, a nightmarish thing intent on scaring him from continuing his dedicated research.

When paranormal activity occurs, when UFOs intrude upon the lives of petrified people, and when researchers of all things paranormal get too close to the truth for their own good, the WIB are ready to strike. They dwell within darkness, they surface when the landscape is black and shadowy, and they spread terror and negativity wherever they walk. Or, on occasion, silently *glide*. They are the Women in Black. Fear them. Keep away from them. And never, *ever*, let them in the house.

CHAPTER 37

CONCLUSIONS

N ow it's time for me to come to some conclusions on the matter of time travel. First and foremost, just about everyone has heard of time travel, even if only in pop culture and entertainment. And there's no doubt that pop culture and entertainment have molded the way we see the phenomenon. That does not, however, change the fact that time travel in the real world appears to be a reality. In many respects, it doesn't seem to operate as we expect it to. For example, while scientists have made claims about being able to travel in time—via black holes and wormholes, for example—very few scientists, if any, have actually come up with a viable way to create a "nuts and bolts" time machine. Indeed, in the pages of this book we see very little about time machines. There's a very good reason for that.

While we have seen and heard numerous stories of time travel in the real world, what we rarely see, if ever, is the vehicle in which the time travelers move across the centuries. One could say that the Roswell device of July 1947 and the craft that came down at Rendlesham Forest in December 1980 are such examples. The designation could also apply to the so-called Flying Triangle type of UFO, which researcher Omar Fowler suspected could be a time machine. And maybe they were exactly that. In so many cases, however, the time traveling appears to occur wholly at random.

In other words, people are suddenly thrown into environments completely unknown. One such instance was the famous case of time traveling at Versailles, France, on August 10, 1901, involving Charlotte Anne Moberly and Eleanor Jourdain. They had no time machine in which to traverse the years in a moment or two. They weren't even looking for a mysterious event. Yet they seem to have found themselves suddenly in a very different timeline. It's important to note that Jourdain and Moberly could see the people of that era, and more intriguing, they could see her. That strongly suggests that time travel is not a one-way situation. It suggests that time travel allows us to interact with people of other times—something that could be incredibly dicey should a person deliberately or accidentally kill their own grandfather in an earlier time, thus ensuring they wouldn't be born. Indeed, such a predicament is called the "grandfather paradox."

While we may see and hear numerous stories of time travel in the real world, we rarely see the vehicle in which the time travelers move across the centuries. Here is one artist's conception of such a craft.

Of this mind-boggling phenomenon, Greg Uyeno says at the website *Space.com*: "The grandfather paradox is a potential logical problem that would arise if a person were to travel to a past time. The name comes from the idea that if a person travels to a time before their grandfather had children, and kills him, it would make their own birth impossible. So, if time travel is possible, it somehow must avoid such a contradiction."

Uyeno continues: "The logical inconsistency of time travel is a common theme in time-warping fiction, but it's also of interest to philosophers. In early versions of the grandfather paradox, some tried to argue that time travel was impossible on logical grounds, said Tim Maudlin, a philosopher at New York University, who frequently writes about physics and philosophy. 'In a way, that's like asking why, right now, I can't be wet and completely dry,' he said. 'Well, that's just logically impossible. What are you asking about?'"

There's no real answer to this particular paradox. Maybe killing your grandfather in an earlier time will create a new timeline. If it does, you will remain alive and healthy in your own, original timeline. Granted, such a thing is theoretical and nothing more.

We should not forget that animals have seemingly crossed through time. The stories in the pages of this book of a pterodactyl, of a mammoth, and of a saber-toothed tiger (Smilodon) all leaping the millennia suggest that much of time traveling is indeed achieved at random. Of course, the problem with the random angle is that we cannot predict when or where we might be able to leap through a "time tunnel," whether for a brief period or forever.

On the other hand, maybe someone else *has* managed to control time. Let us now address the matter of UFO phenomena, which include sightings of UFOs and warnings and threats from the dreaded Men in Black. Both phenomena can be seen from the perspective of a direct connection to time travel and to time travelers. In view of that, the day may very well come when we will have to radically alter our beliefs on the UFO enigma and our long-standing perceptions on what UFOs and their pilots are. Maybe there are no aliens. Maybe, instead, there is an incredible body of time travelers—from our future and maybe from our past—who are masquerading and hiding their true identities. Maybe they are really us and not creatures from other planets or galaxies.

The most disturbing angle of all this might be the possibility that time travelers from a faraway future are using us as cattle.

The most disturbing angle of all this might be the possibility that time travelers from a faraway future are using us as cattle. As I noted in the chapter "Timelines and the Roswell Incident," Jim Penniston "stated that the presumed aliens are, in reality, visitors from a far-flung future. Our future. That future, Penniston added, is very dark, in infinitely deep trouble, and polluted. In it, the human race is overwhelmingly blighted by reproductive problems. The answer to those massive problems, Penniston was told by the entities he met in the woods, was for the entities to travel into the distant past—to our present day—to secure sperm, eggs, and chromosomes in an effort to ensure the continuation of the severely waning human race of tomorrow."

In light of that grim scenario, perhaps the time travelers among us are deliberately presenting themselves as aliens. It would be the perfect camouflage. That may be the reason why they drive around in black cars; perhaps they are doing their best to blend in—but, most of the time, not doing such a good job of it.

Perhaps the time travelers among us are deliberately presenting themselves as aliens. It would be the perfect camouflage.

There's one final issue we should never forget. Throughout this book are many examples of how people like you and me have had brief viewings of the future, albeit in a dream state. Consider the numerous reports of prophetic dreams of nuclear war and of the terrible disaster that killed so many children at Aberfan, Wales. There's little doubt this suggests we already have the ability to travel time—or at least to peer into the future. The problem, however, is that we don't know to harness such an incredible ability. Should the day come when we can harness time, then it will be a whole different ball game.

Finally, there's a possibility that the people of the future are determined to keep us—a warlike, dangerous, and destructive race—from intruding into their time.

TIME TRAVEL: THE SCIENCE AND SCIENCE FICTION

Or, as Michael J. Fox's character Marty McFly said in *Back to the Future*: "I guess you guys aren't ready for that." Perhaps we aren't. Worse still, remember the words of Dr. Zaius to astronaut Taylor in the time travel–themed 1968 movie *Planet of the Apes*: "Don't look for it, Taylor. You may not like what you find." Wise words? Time (ahem) may tell.

FURTHER READING

"Aberfan Disaster: First Photographer on Scene Retires." September 19, 2020. https://www.bbc.com/news/uk-wales-54152447.

"Aberfan Disaster: Teacher Who Rescued Pupils Dies Aged 86." May 8, 2020. https://www.bbc.com/news/uk-wales-52593473.

Adams, Tom. *The Choppers—and the Choppers: Mystery Helicopters and Animal Mutilations.* Paris, TX: Project Stigma, 1991.

Adamski, George. *Behind the Flying Saucer Mystery.* New York: Paperback Library Edition, 1967.

"Al Bielek—The Man Who Traveled through Time and Space." 2021. https://ocg hostsandlegends.com/al-bielek-the-man-who-traveled-through-time-and-space/.

"Alleged Secret Underground Alien Base in Dulce, New Mexico, The." *Curiosity Makes You Smarter.* July 30, 2019. https://curiosity.com/topics/the-alleged-secret-underground-alien-base-in-dulce-new-mexico-curiosity/.

Allen-Block, Irene. *The Psychic Spy.* Wales: Glannant Ty Media, 2013.

Andrews, Bill. "If Wormholes Exist, Could We Really Travel through Them?" July 30, 2019. https://www.discovermagazine.com/the-sciences/if-wormholes-exist-could-we-really-travel-through-them.

Angelucci, Orfeo. *The Secret of the Saucers.* Stevens Point, WI: Amherst Press, 1955.

"Animal Mutilation." *The Vault.* 2017. https://vault.fbi.gov/Animal%20Mutilation.

"Arizona, Thunderbird, The." *Weird U.S.* 2021. http://www.weirdus.com/states/arizona/bizarre_beasts/thunderbird/index.php.

Ausiello, Michael. "The Sept. 11 Parallel 'Nobody Noticed' ('Lone Gunmen' Pilot Episode Video)." June 21, 2002. http://www.freerepublic.com/focus/news/703915/posts.

TIME TRAVEL: THE SCIENCE AND SCIENCE FICTION

Austin, Jon. "Crop Circles Are 'Messages from Future Humans That Prove Time Travel.'" March 12, 2017. https://www.express.co.uk/news/weird/798160/Crop-circles-messages-future-humans-prove-time-travel-aliens-UFO.

Banias, M. J. "Chicago's Current Mothman Flap 'A Warning,' Says Expert." June 7, 2017. http://mysteriousuniverse.org/2017/06/chicagos-current-mothman-flap-a-warning-says-expert.

Barker, Gray. *M.I.B.: The Secret Terror among Us.* Clarksburg, WV: New Age Press, 1983.

———. *They Knew Too Much about Flying Saucers.* New York: University Books, 1956.

———. *The Strange Case of Dr. M. K. Jessup.* Point Pleasant, WV: New Saucerian Books, 2014.

Beckley, Timothy. *The UFO Silencers: Mystery of the Men in Black.* New Brunswick, NJ: Inner Light Publications, 1990.

———. *The UFO Silencers (Special Edition).* New Brunswick, NJ: Inner Light Publications, 1990.

Beckley, Timothy Green, and John Stuart. *Curse of the Men in Black.* New Brunswick, NJ: Global Communications, 2010.

Beckley, Timothy, and Christa Tilton. *Underground Bio Lab at Dulce: The Bennewitz UFO Papers.* New Brunswick, NJ: Inner Light Productions, 2012.

Bender, Albert. *Flying Saucers and the Three Men.* New York: Paperback Library, 1968.

Bennett, Colin. *Looking for Orthon: The Story of George Adamski, the First Flying Saucer Contactee, and How He Changed the World.* New York: Paraview Press, 2001.

Bennett-Smith, Meredith. "Mystery of 1938 'Time Traveler' with Cell Phone Solved?" April 4, 2013. http://www.huffingtonpost.com/2013/04/04/time-traveler-cell-phone-1938-video-woman-factory_n_3013996.html.

Berlitz, Charles, and William Moore. *The Philadelphia Experiment: The True Story Behind Project Invisibility.* St. Albans, U.K.: Granada Publishing, 1980.

Berry, C. R. "Were the Green Children of Woolpit Time Travellers from a Dystopian Earth?" December 13, 2018. http://timetravelnexus.com/were-the-green-children-of-woolpit-time-travellers-from-a-dystopian-earth/.

"Binary Code." October 31, 2018. https://www.techopedia.com/definition/17052/binary-code.

Bishop, Greg. *Project Beta: The Story of Paul Bennewitz, National Security, and the Creation of a Modern UFO Myth.* New York: Paraview–Pocket Books, 2005.

Bishop, Jason III. "The Dulce Base." *Sacred Texts.* 2017. http://www.sacred-texts .com/ufo/dulce.htm.

Bommersmach, Jana. "Tombstone's Flying Monster." June 1, 2007. https://truewest magazine.com/tombstone-epitaph/.

Branton. "The Dulce Book." *Whale.* 2017. http://www.whale.to/b/dulce_b.html.

Bratschi, Pierre. "Black Holes May Make Time Travel Possible." December 18, 2017. https://www.horizons-mag.ch/2017/12/18/black-holes-may-make-time-travel-possible/.

Bruce, Alexandra. *The Philadelphia Experiment Murder.* New York: Sky Books, 2001.

Carlton, Genevieve. "Everything You Should Know about the Moberly-Jourdain Incident." September 13, 2019. https://www.ranker.com/list/moberly-jourdain-incident-and-time-travel/genevieve-carlton.

Central Intelligence Agency. "Mars Exploration, May 22, 1984." 2018. https://www .cia.gov/library/readingroom/document/cia-rdp96-00788r001900760001-9.

"Chaplin's Time Traveler." October 19, 2010. https://www.youtube.com/watch?v =Y6a4T2tJasU.

Cofield, Calla. "Time Travel and Wormholes: Physicist Kip Thorne's Wildest Theories." December 19, 2014. https://www.space.com/28000-physicist-kip-thorne-wildest-theories.html.

Coleman, Loren. *Mothman and Other Curious Encounters.* New York: Paraview Press, 2002.

Collins, Andrew. *The New Circlemakers: Insights into the Crop Circle Mystery.* Virginia Beach, VA: A.R.E. Press/4th Dimension Press, 2009.

Corbett, James. "Lone Gunmen Producer Questions Government on 9/11." February 25, 2008. http://www.corbettreport.com/articles/20080225_gunmen_911.htm.

Corso, Philip J., and William J. Birnes. *The Day after Roswell.* New York: Simon & Schuster, 1997.

Cunningham, Gary, and Ronan Coghlan. *The Mystery Animals of Ireland.* Woolsery, U.K.: CFZ Press, 2010.

Cutchin, Joshua. *A Trojan Feast.* San Antonio, TX: Anomalist Books, 2015.

TIME TRAVEL: THE SCIENCE AND SCIENCE FICTION

Davis, Eric W. *Teleportation Physics Study*. August 2004. https://fas.org/sgp/eprint/tele port.pdf.

Deem, James M. "The Aberfan Tragedy: A Precognitive Investigation." 2020. https://jamesmdeem.com/stories.time.aberfan.html.

Devlin, Hannah. "Beam Me Up, Scotty! Scientists Teleport Photons 300 Miles into Space." July 12, 2017. https://www.theguardian.com/science/2017/jul/12/ scotty-can-you-beam-me-up-scientists-teleport-photons-300-miles-into-space.

"Did the Philadelphia Experiment Really Happen?" November 19, 2019. https:// www.gaia.com/article/has-the-government-already-achieved-time-travel.

"Doctor Who." 2021. https://www.doctorwho.tv/.

"Doppelgänger." *Britannica*. 2021. https://www.britannica.com/art/doppelganger.

"Doppelgänger." *Paranormal Guide*. October 4, 2013. http://www.theparanormal guide.com/blog/doppelganger.

Ebert, Roger. "It's a Wonderful Life." January 1, 1999. https://www.rogerebert.com/ reviews/great-movie-its-a-wonderful-life-1946.

———. "Not Your Average Los Angeles Cult." 2012. https://www.rogerebert.com/ reviews/the-sound-of-my-voice-2012.

Eldritch, Tristan. "The Green Children of Woolpit." 2011. http://2012diaries.blog spot.com/2011/12/green-children-of-woolpit.html.

"11:11: Is It Happening to You?" 2021. https://www.powerofpositivity.com/1111-is-it-happening-to-you/.

Farrell, Joseph P. *Roswell and the Reich*. Kempton, IL: Adventures Unlimited Press, 2010.

"Feel Like You've Been Here Before? It Might Be Déjà Vu." July 2, 2019. https:// www.pennmedicine.org/updates/blogs/health-and-wellness/2019/july/deja-vu.

Firth, Niall. "Is This a Time-Traveler in a Charlie Chaplin Film? Footage from 1928 Shows Woman 'Using a Mobile Phone.'" November 1, 2010. http://www.daily mail.co.uk/sciencetech/article-1324132/Time-traveller-woman-mobile-phone-1928-Charlie-Chaplin-film.html.

Fitzgerald, Paul. "Time Traveling Brothers Reveal What the Distant Future Will Be Like." September 26, 2019. https://www.thetorontotribune.com/2017/09/ 26/time-traveling-brothers-reveal-distant-future-will-like/.

"Folklore of Bhutan—Migoi, the Yeti." 2021. https://bhutancanada.org/folklore-of-bhutan-migoi-the-yeti/.

Friedman, Stanton T., and Don Berliner. *Crash at Corona.* New York: Paragon House, 1992.

Fuller, John G. *The Interrupted Journey: Two Hours "Aboard a Flying Saucer."* New York: Dial Press, 1966.

"'Future' in Flux: Maybe We Don't Have the Capacitor to Change.'" 2020. http://www.widescreenings.com/back-future-analysis-review.html.

Galde, Phyllis, editor. *The Best of Roswell.* Lakeville, MN: Galde Press, 2007.

"GAO Report on Roswell, NM UFO Crash." July 28, 1995. https://fas.org/sgp/othergov/roswell.html.

General Accounting Office. *Results of a Search for Records Concerning the 1947 Crash Near Roswell, New Mexico.* Washington, D.C.: U.S. Government Printing Office, July 28, 1995.

Gillan, Joanna. "Doppelgängers and Curious Myths and Stories of Spirit Doubles." April 1, 2020. https://www.ancient-origins.net/myths-legends/doppelgangers-and-mythology-spirit-doubles-001825.

Goldberg, Dr. Bruce. *Time Travelers from Our Future.* St. Paul, Minnesota: Llewellyn Publications, 1998.

Gonzalez, Dave. "Inside the Real-Life Time-Travel Experiment That Inspired 'Stranger Things.'" August 30, 2016. https://www.thrillist.com/entertainment/nation/stranger-things-true-story-montauk-project-philadelphia-experiment.

Gorvett, Zaria. "You Are Surprisingly Likely to Have a Living Doppelgänger." July 13, 2016. https://www.bbc.com/future/article/20160712-you-are-surprisingly-likely-to-have-a-living-doppelganger.

"Green Children of Woolpit, The." 2014. http://www.mysteriousbritain.co.uk/england/suffolk/folklore/the-green-children-of-woolpit.html.

Griffin, Andrew W. "Riders on the Storm (Strange Days Have Tracked Us Down)." November 10, 2016. http://www.reddirtreport.com/red-dirt-grit/riders-storm-strange-days-have-tracked-us-down.

Guiley, Rosemary Ellen. *Slips in Time and Space.* New Milford, CT: Visionary Living Publishing, 2019.

TIME TRAVEL: THE SCIENCE AND SCIENCE FICTION

Halpenny, Bruce Barrymore. *Ghost Stations IV: True Ghost Stories*. U.K.: Caldec, 1991.

Hanks, Micah. "The Electronic Fog: A Strange Trip through Time, Space, and Tailwinds." July 24, 2015. https://mysteriousuniverse.org/2015/07/the-electronic-fog-a-strange-trip-through-time-space-and-tailwinds/.

Harpur, Merrily. *Mystery Big Cats*. Wymeswold, U.K.: Heart of Albion Press, 2006.

Haughton, Brian. "The Green Children of Woolpit." 2010. http://brian-haughton .com/articles/green-children-of-woolpit/.

Holdstock, Robert. *Mythago Wood*. New York: Arbir House, 1984.

Holiday, F. W. *The Goblin Universe*. St. Paul, MN: Llewellyn Publications, 1986.

Hollis, Heidi. *The Hat Man: The True Story of Evil Encounters*. Milwaukee, WI: Level Head, 2014.

Holman, Richard F. "Follow Up: What Exactly Is a 'Wormhole'? Have Wormholes Been Proven to Exist or Are They Still Theoretical?" September 15, 1997. https://www.scientificamerican.com/article/follow-up-what-exactly-is-/.

"Horseman of Nottlebrush Down, The." September 16, 2018. https://anomalien .com/the-horseman-of-bottlebrush-down/.

Howell, Elizabeth. "Time Travel: Theories, Paradoxes & Possibilities." November 14, 2017. https://www.space.com/21675-time-travel.html.

Invisible Man, The. 2000–02. https://www.imdb.com/title/tt0220238/.

Jacobsen, Annie. *Area 51: An Uncensored History of America's Top Secret Military Base*. New York: Little, Brown, 2012.

Jones, Marie D., and Larry Flaxman. *11:11, The Time Prompt Phenomenon*. Newburyport, MA: New Page Books, 2019.

———. *This Book is from the Future*. Pompton Plains, NJ: New Page Books, 2012.

Keel, John. *The Mothman Prophecies*. New York: Tor Books, 1991.

Keith, Jim. *Black Helicopters over America: Strikeforce for the New World Order*. Lilburn, GA: IllumiNet Press, 1994.

———. *Black Helicopters II: The Endgame Strategy*. Lilburn, GA: IllumiNet Press, 1997.

Kennedy, Sequoyah. "Amazon Alexa Says Chemtrails Are a Real Government Conspiracy." April 23, 2018. http://mysteriousuniverse.org/2018/04/amazon-alexa-says-chemtrails-are-a-real-government-conspiracy/.

TIME TRAVEL: THE SCIENCE AND SCIENCE FICTION

Kerner, Nigel. *Grey Aliens and the Harvesting of Souls: The Conspiracy to Genetically Tamper with Humanity*. Rochester, VT: Bear & Company, 2010.

———. *The Song of the Greys*. London: Hodder & Stoughton, 1997.

Khanna, Gaurav. "Maybe You Really Can Use Black Holes to Travel the Universe." January 25, 2019. https://www.discovermagazine.com/the-sciences/maybe-you-really-can-use-black-holes-to-travel-the-universe.

Knapp, Francky. "The Accidental Time Travel Incident with Marie Antoinette." January 16, 2018. https://www.messynessychic.com/2018/01/16/the-accidental-time-travel-incident-with-marie-antoinette/.

Knapton, Sarah. "Dog Walker Met UFO 'Alien' with Scandinavian Accent." March 22, 2009. https://www.telegraph.co.uk/news/newstopics/howaboutthat/5031587/Dog-walker-met-UFO-alien-with-Scandinavian-accent.html.

Koslovic, Melanie. "Green Children of Woolpit." July 16, 2014. http://prezi.com/re8juu0kv0eb/green-children-of-woolpit/.

laymwe01. "Green Children of Woolpit: Mysterious Visitors from an Unknown Land." 2014. http://laymwe01.hubpages.com/hub/Green-Children-of-Woolpit-Mysterious-visitors-from-an-unknown-land.

Leung, Clint. "Thunderbird: A Symbol of Power, Strength and Nobility." 2020. https://www.cuyamungueinstitute.com/articles-and-news/thunderbird-a-symbol-of-power-strength-and-nobility/.

Little, Tom. "Tracing the Development of the Doppelgänger." November 6, 2017. https://www.atlasobscura.com/articles/history-doppelganger.

Lynch, Patrick. "10 Premonitions of Doom from History That Actually Came True." 2021. https://historycollection.com/10-premonitions-of-doom-and-from-history-that-actually-came-true/.

Lyons, Gordon. "Vickers Wellington N2980, Loch Ness Highland." 2015. http://www.aircrashsites-scotland.co.uk/wellington_loch-ness.htm.

MacKenzie, Andrew. *Adventures in Time: Encounters in the Past*. London: Athlone Press, 1997.

Marrs, Jim. *PSI Spies: The True Story of America's Psychic Warfare Program*. Newburyport, MA: Weiser Books, 2007.

McGowan, Kathleen. "The Mystery of the Versailles Time Slip." 2013. http://www.kathleenmcgowan.com/the-mystery-of-the-versailles-time-slip/.

TIME TRAVEL: THE SCIENCE AND SCIENCE FICTION

McNeill, Leila. "The Edwardian Women Who Claimed to Travel Back in Time." August 22, 2018. https://www.mentalfloss.com/article/554109/edwardian-women-who-claimed-travel-back-time.

Medway, Gareth J. "MIB Encounters: Men in Black Encounters, a Short Catalogue." 2021. https://pelicanist.blogspot.com/p/mib-encounters.html.

"Military Mayhem in Montauk, Long Island." 2021. http://www.weirdus.com/states/new_york/unexplained_phenomena/montauk_project/index.php.

"Millennium." 2021. https://www.amazon.com/Millennium-Kris-Kristofferson/dp/B00005O5AW.

Miller, Liz Shannon. "'Time after Time' Review: A Super-Hot Jack the Ripper Evokes Peculiar Questions in Sporadically Fun Time Travel Drama." March 3, 2017. https://www.indiewire.com/2017/03/time-after-time-review-tv-show-jack-the-ripper-hot-time-travel-1201789414/.

"Missing Thunderbird Photo, The." 2020. https://themothman.fandom.com/wiki/The_Missing_Thunderbird_Photo.

Mosher, Dave. "Nuclear Bombs Trigger a Strange Effect That Can Fry Your Electronics—Here's How It Works." June 7, 2017. http://www.businessinsider.com/nukes-electromagnetic-pulse-electronics-2017-5.

Naish, Darren. "What Was the Montauk Monster?" August 3, 2008. https://scienceblogs.com/tetrapodzoology/2008/08/04/the-montauk-monster.

"Native American Bigfoot Figures of Myth and Legend." 2020. http://www.native-languages.org/legends-bigfoot.htm.

"Native American Legends: Thunderbird (Thunder-Birds)." 2020. http://www.native-languages.org/thunderbird.htm.

Newman, Bernard. *The Flying Saucer*. Yardley, PA: Westholme Publishing, 2010.

Nickell, Joe. "Montauk Monster and the Racoon Body Farm." November 2, 2012. https://skepticalinquirer.org/newsletter/montauk-monster-and-the-raccoon-body-farm/.

O'Brien, Christopher. *Stalking the Herd: Unraveling the Cattle Mutilation Mystery*. Kempton, IL: Adventures Unlimited Press, 2014.

O'Connell, Cathal. "Five Ways to Travel through Time." April 5, 2016. https://cosmosmagazine.com/science/physics/five-ways-to-travel-through-time/.

Offutt, Jason. "A Case for Time Travelers." October 8, 2017. https://mysterious universe.org/2017/10/a-case-for-time-travelers/.

———. *Darkness Walks: The Shadow People among Us*. San Antonio, TX: Anomalist Books, 2009.

"Pennine Pterodactyl." 2021. https://obscurban-legend.fandom.com/wiki/ Pennine_Pterodactyl.

Penniston, Jim, and Gary Osborn. *The Rendlesham Enigma*. Independently published, July 22, 2019.

"Planet of the Apes." 2020. https://planetoftheapes.fandom.com/wiki/Planet_of_ the_Apes_Wiki.

"'Project Gasbuggy' Atomic Explosion Site." 2017. https://www.roadsideamerica .com/story/16912.

"Quantum Teleportation." 2021. https://researcher.watson.ibm.com/researcher/ view_group.php?id=2862.

Radford, Benjamin. "The Crop Circle Mystery: A Closer Look." June 10, 2017. https://www.livescience.com/26540-crop-circles.html.

Ramsland, Katherine. "Shadow People." *Psychology Today*, July 14, 2013. https://www .psychologytoday.com/us/blog/shadow-boxing/201307/shadow-people.

Randle, Kevin D. *The Roswell Encyclopedia*. New York: HarperCollins, 2000.

Randle, Kevin D., and Donald R. Schmitt. *UFO Crash at Roswell*. New York: Avon Books, 1991.

Randles, Jenny. *Breaking the Time Barrier*. New York: Paraview Pocket Books, 2005.

———. *The Truth Behind the Men in Black*. London: Piatkus, 1997.

———. *Time Storms: The Amazing Evidence of Time Warps, Space Rifts and Time Travel*. London: Piatkus, 2001.

Redd, Nola Taylor. "How Fast Does Light Travel?" March 7, 2018. https://www .space.com/15830-light-speed.html.

———. "What Is Wormhole Theory?" October 21, 2017. https://www.space.com/ 20881-wormholes.html.

TIME TRAVEL: THE SCIENCE AND SCIENCE FICTION

Redfern, Nick. *The Bigfoot Book: The Encyclopedia of Sasquatch, Yeti, and Cryptid Primates.* Canton, MI: Visible Ink Press, 2015.

———. *The Black Diary.* Bracey, VA: Lisa Hagan Books, 2018.

———. "Body Snatchers: Before and Beyond." In *Darklore Volume III*, edited by Greg Taylor. Brisbane, Australia: Daily Grail Publishing, 2009.

———. *Body Snatchers in the Desert.* New York: Simon & Schuster, 2005.

———. *Final Events: And the Secret Government Group on Demonic UFOs and the Afterlife.* San Antonio, TX: Anomalist Books, 2010.

———. Interview with Jenny Burrows, January 23, 2009.

———. Interview with Jill O'Brien, January 9. 2009.

———. Interview with Joshua P. Warren, October 6, 2010.

———. Interview with Mac Tonnies, March 14, 2004.

———. Interview with Mac Tonnies, September 9, 2006.

———. Interview with Mac Tonnies, July 7, 2009.

———. Interview with Matthew Williams, October 3, 2006.

———. Interview with Neil Arnold, April 4, 2012.

———. Interview with Nick Pope, January 22, 1997.

———. Interview with Pauline Charlesworth, July 16, 1997.

———. Interview with Ronan Coghlan, March 18, 2012.

———. Interview with Susan Sheppard, August 28, 2017.

———. Interview with Tracie Austin, September 21, 2017.

———. *Men in Black.* Bracey, VA: Lisa Hagan Books, 2015.

———. *Nessie: Exploring the Supernatural Origins of the Loch Ness Monster.* Bracey, VA: Lisa Hagan Books, 2018.

———. *The Rendlesham Forest UFO Conspiracy: A Close Encounter Exposed as a Top Secret Government Experiment.* Bracey, VA: Lisa Hagan Books, 2020.

TIME TRAVEL: THE SCIENCE AND SCIENCE FICTION

———. "A Response from Nick Redfern." *Fate*, September 2005.

———. *The Roswell UFO Conspiracy: Exposing a Shocking and Sinister Secret*. Bracey, VA: Lisa Hagan Books, 2017.

———. *Women in Black*. Bracey, VA: Lisa Hagan Books, 2016.

Ridpath, Ian. "Rendlesham Forest UFO—The Witness Statements." January 2021. http://www.ianridpath.com/ufo/rendlesham2b.html.

Riesman, Abraham. "*12 Monkeys* Is the Apocalypse Movie We Need Right Now." March 13, 2020. https://www.vulture.com/2020/03/12-monkeys-why-terry-gilliams-movie-is-so-relevant-today.html.

"Rollright Stones, The." 2021. https://www.rollrightstones.co.uk/.

Rose, Kate. "The Phenomenon & Meaning of 11:11: The Twin Flame Connection." June 5, 2015. https://www.elephantjournal.com/2015/06/the-phenomenon-meaning-of-1111-the-twin-flame-connection/.

Russell, Davy. "Invisible Sasquatch." February, 29, 2000. http://www.bigfoot encounters.com/articles/invisible.htm.

Sample, Ian. "Scientists Use Stem Cells from Frogs to Build First Living Robots." January 13, 2020. https://www.theguardian.com/science/2020/jan/13/scien tists-use-stem-cells-from-frogs-to-build-first-living-robots.

Sandeman, George. "Scientist Claims Crop Circles Are Hidden Messages Left by Aliens or Human Time Travellers." September 13, 2017. https://www.thesun .co.uk/tech/3071298/scientist-claims-crop-circles-are-hidden-messages-left-by-aliens-or-human-time-travellers/.

Sauve, Mike. *Who Authored the John Titor Legend?* Woodland Hills, CA: Big Swerve Press, 2016.

Scharr, Jillian. "Wormhole Is Best Bet for Time Machine, Astrophysicist Says." August 25, 2013. https://www.livescience.com/39159-time-travel-with-worm hole.html.

Seaburn, Paul. "The Mysterious John Titor, Trump and Today's Time Travelers." February 27, 2018. https://mysteriousuniverse.org/2018/02/the-mysterious-john-titor-trump-and-todays-time-travelers/.

———. "Physicist Describes Wormhole That Will Make Time Travel Possible." November 19, 2017. https://mysteriousuniverse.org/2017/11/physicist-describes-wormhole-that-will-make-time-travel-possible/.

TIME TRAVEL: THE SCIENCE AND SCIENCE FICTION

Shuker, Dr. Karl. "The Green Children of Woolpit: Investigating a Medieval Mystery." November 29, 2012. http://eclectariumshuker.blogspot.com/2012/11/the-green-children-of-woolpit.html.

———. "Seeking the Missing Thunderbird Photograph—One of Cryptozoology's Most Tantalizing Unsolved Cases." November 25, 2014. http://karlshuker.blogspot.com/2014/11/seeking-missing-thunderbird-photograph.html.

Simpson, George E., and Neal R. Burger. *Thin Air.* London: New English Library, 1979.

Spender, Tom. "Teleportation: Photon Particles Today, Humans Tomorrow?" BBC. July 14, 2017. https://www.bbc.com/news/science-environment-40594387.

Steiger, Brad. *The Philadelphia Experiment & Other UFO Conspiracies.* New Brunswick, NJ: Inner Light Publications, 1990.

Steinman, Paige. "Claimed Time Traveler Says He Has Dire Predictions about Year 2749." June 15, 2020. https://www.icepop.com/time-traveler-future-predictions-2749/2/.

Stierwalt, Sabrina. "Can Science Explain Déjà Vu?" March 23, 2020. https://www.scientificamerican.com/article/can-science-explain-deja-vu/.

"Street Eyes." 2017. https://www.amazon.com/Street-Eyes-Steve-Helmkamp/dp/B076BVSQJ5.

Strieber, Whitley. *Communion: A True Story.* New York: Beach Tree Books, 1987.

———. *Warday.* New York: Warner Books, 1984.

Stringfield, Leonard. *Situation Red: The UFO Siege.* London: Sphere Books, 1978.

———. *UFO Crash Retrievals: The Inner Sanctum.* Cincinnati, OH: privately published, 1991.

Swancer, Brent. "Bizarre Cases of Real-Life Time Travelers." July 25, 2016. https://mysteriousuniverse.org/2016/07/bizarre-cases-of-real-life-time-travelers/.

Tangermann, Victor. "Paradox-Free Time Travel Possible with Many Parallel Universes." December 13, 2019. https://futurism.com/paradox-free-time-travel-parallel-universes.

Taylor, Greg. "New Roswell Revelations." May 1, 2009. http://dailygrail.com/news/new-roswell-revelations.

Teale, Julia C., and Akira R. O'Connor. "What Is Déjà Vu?" February 27, 2015. https://kids.frontiersin.org/article/10.3389/frym.2015.00001.

"Thunderbird of Native Americans, The." 2020. https://www.legendsofamerica .com/thunderbird-native-american/.

"Time after Time." IMDB. 2021. https://www.imdb.com/title/tt5031234/.

"Time after Time." TCM. 2021. https://www.tcm.com/tcmdb/title/25060/time-after-time#overview.

"Time Machines." American Museum of Natural History. 2021. https://www.amnh .org/exhibitions/einstein/time/time-machines.

"Time Traveler Caught on Film—Time Traveler Cell Phone—Charlie Chaplin Time Traveler." October 27, 2010. https://www.youtube.com/watch?v=TiIr-pEMbQ2M.

Tonnies, Mac. *After the Martian Apocalypse.* New York: Paraview–Pocket Books, 2004.

———. *The Cryptoterrestials.* San Antonio, TX: Anomalist Books, 2010.

Tracey, Janey. "Five Crazy Conspiracy Theories about the Roswell UFO." July 1, 2014. https://www.outerplaces.com/science-fiction/item/4715-top-five-cra ziest-roswell-conspiracy-theories.

———. "Physicist Claims Traveling Back in Time Creates Doppelgängers That An-nihilate Each Other." July 29, 2015. https://www.outerplaces.com/science/ item/9464-physicist-claims-traveling-back-in-time-creates-doppelgangers-that-annihilate-each-other.

United States Air Force. *The Roswell Report: Case Closed.* U.S. Government Printing Office, 1997.

United States Air Force. *The Roswell Report: Fact versus Fiction in the New Mexico Desert.* U.S. Government Printing Office, 1994.

Wargo, Eric. *Time Loops: Precognition, Retrocausation, and the Unconscious.* San Antonio, TX: Anomalist Books, 2018.

Warren, Joshua P. *Pet Ghosts: Animal Encounters from Beyond the Grave.* Franklin Lakes, NJ: New Page Books, 2006.

Weatherly, David. *Black Eyed Children.* Denton, TX: Leprechaun Press, 2012.

———. *Strange Intruders.* Denton, TX: Leprechaun Press, 2013.

"Welcome to Woolpit Village." 2014. http://www.woolpit.org/.

"Western Electric Model 34A 'Audiphone' Carbon Hearing Aid." 2015. http://www .hearingaidmuseum.com/gallery/Carbon/WesternElectric/info/westelect34a .htm.

"What Causes Déjà Vu?" 2021. https://www.healthline.com/health/mental-health/ what-causes-deja-vu.

"What Is a Sabertooth?" 2021. https://ucmp.berkeley.edu/mammal/carnivora/ sabretooth.html.

Whyte, Chelsea. "Time Travel without Paradoxes Is Possible with Many Parallel Timelines." December 13, 2019. https://www.newscientist.com/article/222 7304-time-travel-without-paradoxes-is-possible-with-many-parallel-timelines/.

Wild, Flint. "What Is a Black Hole?" August 21, 2018. https://www.nasa.gov/ audience/forstudents/k-4/stories/nasa-knows/what-is-a-black-hole-k4.html.

Williams, Robert. "'A Christmas Carol' with Alistair Sim (1951)." December 9, 2010. http://robertwilliamsofbrooklyn.blogspot.com/2010/12/christmas-carol- with-alistair-sim-1951.html.

X, Commander. *The Philadelphia Experiment Chronicles*. Wilmington, DE: Abelard Pro- ductions, 1994.

INDEX

NOTE: (ILL) INDICATES PHOTOS AND ILLUSTRATIONS